Direct to Brain Windows 2.0

What They Never Show Us

Ted Huntington

Direct to Brain Windows 2.0
What They Never Show Us

ISBN (978-0-9881922-4-9)

www.tedhuntington.org

Dedicated to the many people excluded from direct-to-brain windows and victims of remote particle nanotechnology everywhere (if such things exist)

Contents

Introduction

Is "neuron reading and writing" an actual science? Yes, of course it is! And so are "remote neuron reading and writing", "recording thought-sounds", and "recording thought-images". But why is there only silence about what is obviously a very important field of science? Are any governments, companies, and universities working on remote neuron reading and writing? If yes, why don't they call it "remote neuron reading and writing"? Why have we never heard of "remote neuron reading and writing" before? Could it be that there is a very strict and harshly punished restraint on talking openly about this very important and ancient science by people who perfected it centuries earlier and now trickle it out in crumbs to the many poor D2BW consumers in return for many millions of dollars?

Here are more details to help excluded people understand and defend themselves against the many vicious remote neuron reading and writing owners and consumers.

What does it look like to see direct-to-brain windows?

As an excluded I don't know for sure what receiving D2B looks like, but it seems pretty clear that it looks similar to a computer desktop (people have many windows open) – but the windows are in front of your eyes and are probably semi-transparent. But with D2B, windows can be anywhere in a person's view. There could be giant windows. A person could be surrounded by a cylinder of square tiled windows. Mostly people probably just have square and circle two dimensional (flat) windows. Most of the windows are probably semi-transparent, but a person could see a movie with "total immersion", that is, that the movie completely and opaquely fills their eye screen.

Possibly people may view windows on their thought-screen too.

What kind of direct-to-brain windows do people get?
I don't know for sure, but it seems likely that some common D2B windows are: 1) Any window you might have on your computer, like a web browser, email, instant messages, Word Processor, Video Editor, remote desktop windows to their home computers, calculator, a clock that shows the date and time (wouldn't it be so nice to be able to quickly popup the time in front of your eyes to know if you need to hurry or have time to do some activity?), a calendar (just like many people have a paper calendar, and now electronic calendars to organize their time, and to match open times with other people in the future). The memory that stores all the events on the calendars is, presumably, at the neuron D2B provider company, which is presumably the undemocratic monopoly of AT&T and Verizon, here in the USA. There must be tight controls on the D2B data. There are obviously no videos of people

and their thought-screens and thought-audio captured by microscopic cameras that have reached the public. Probably average D2B consumers cannot store D2B data in a computer in their home, the way the Internet is. Where is all the D2B data stored? 2) Video windows with a live image of your mate, kids, friends, and co-workers. These windows may stay on constantly or only appear very frequently. Since the human mind can only really focus on a few things at a time, perhaps there are just frequent updates, like instant video messages, that D2B consumers just notice and don't think much about. Clearly, many D2B consumers watch and are fixated on their enemies, judging from the large quantity of remote molestation and nasty shills that I endure. 3) Video windows of those around you: It seems likely that as a D2B consumer walks around, windows of people around them may pop up showing a quick profile of the person, and relevant videos from their past. 4) There must be back logged video-messages just like email for people that D2B consumers periodically look at. 5) Pop up "money to you" windows. These windows must offer to pay the D2B consumer money (probably in the form of "virtual money" used to pay for more D2B services. The offers are to pay money to D2B consumers to "shill", that is, to say things, usually unpleasant things, to excluded people, for small jobs like even just to walk some where, to walk past, or walk with an excluded person, money offers for dates, and to perform pleasure-based (sex) services. Money, like a big rotating virtual 3-D "$50" sign might appear in distant 3D locations as seen by the D2B consumer, the D2B consumer then goes to that location, and the $50 sign probably disappears with a sound like "ching!", and enters the D2B consumers account; kind of like a 3D computer game, or leading ducks with bread crumbs, but with humans and money. There may be little virtual money signs on people's bodies, like on their lips for a kiss, etc. It's probably much easier to direct people's motions with little D2B virtual signs.

6) Pop up "money from you" windows: Just like there are windows that offer to pay the D2B consumers for various things, so there must be windows that offer to sell things to D2B consumers in exchange for money. This may include offers for products or services that you specifically may be currently looking to buy, and offers to pay-to-view popular videos (and videos apparently, and fortunately, with very little or no restrictions on privacy, sexual or violent content). 7) Popular videos. D2B service may cost a monthly fee, like a phone bill, but some videos may be free to view. Clearly some videos must cost money to view. Like news stories, there must be tons of videos of recent events on Earth to choose from. Probably the most watched videos are of sex and violence. Much of these interesting videos are probably funded by the wealthiest people on Earth. For example to make people have sex or do violence it must cost wealthy D2B consumers thousands of dollars. Many of the videos must be of excluded people who bit on some bad remote neuron suggestion and did something stupid that many of the D2B consumers then pay to see. You can see how the money that people pay to see popular videos funds the making of more similar videos. D2B consumers apparently see many videos from inside people's houses. Some of these videos may be free, but probably many cost money. Probably there is a constant stream of "death of person A", or "sex between A and B" video offers. Videos of murders, suicides, accidents, crimes (including sex crimes- often produced by remote neuron writing), people nude or having sex in their houses, etc. are probably all advertised and available to those who receive D2BW. Clearly D2BW consumers are not punished by the same laws the excluded public is punished by relating to privacy and pornography. In particular people probably pay to watch videos of the deaths of famous people, and famous crimes from the past (even the far past- hundreds of years ago), to know who killed who- in particular to see from the eyes and to hear the

thought-audio of all involved. Probably many people are interested to see videos about crimes that involved somebody they cared about in order to know the truth about what really happened. Excluded people have to piece together the tiny fragments of evidence available to the public, but D2B consumers probably have most questions answered about who is responsible for any given crime. Many people probably want to see videos of their ancestors and their thoughts as far back as the secret videos go. There is probably a lot of watching the "most beautiful" and "most interesting" people. Many D2B consumers are clearly interested in seeing those excluded who know about D2B (all 3 of us) and in particular that are publicly telling the truth about D2B. In terms of what videos D2B consumers see, there is a parallel, although extremely less restricted, with the Internet. So, for example, if you are a person who dreams of moving to the Moon or Mars, you probably want the latest videos about the forefront of that effort, etc. 8) Possibly some kind of Neuron company administrator people popup windows: It may be that the neuron administrators popup to communicate, or possibly even just text messages to the eyes of D2B consumers- many times apparently warnings, teleprompting, or perhaps coaching what to say out loud and/or what to do. 9) Remote neuron drawings that are made to convey some message. Many times, even as a D2B excluded, I get little tiny visual (and certainly many annoying audio) neuron writings. One example is how a little white dot goes down, and this may be sent to make the receiver think that something they have thought or done has lowered their popularity or is receiving disapproval among D2B owners and/or consumers.

The "Eye-Screen"

Unlike windows on your computer, the windows of D2B are probably mostly of people, and all people probably have two circular or oval screens, over, or around their head (like Mickey Mouse ears): their "eye screen" and their "thought screen". You may hear D2B consumers talking about seeing "eyes". They are talking about seeing what people's eyes see. When you can remotely read what the eye neurons see, people become like walking cameras. The eye screen shows what the brain's eyes see. If you are looking directly in the eyes of any organism with eyes, seeing their eye screen is like looking in a mirror because you see what they see, which is you. We tend to view the image our eyes see with a lot of depth, but the image in our eyes is a two dimensional image that can be copied and sent around like any JPEG image. There is no need to take a photo of something anymore, because we only need to look at it. But, the current problem is that none of our past or present eye images are available for us to access. So we still need to capture images with our own cameras if we ever want to see the image again.

You can imagine that images sent to the eye screen could be used to trick a person into thinking that some object is there when, in reality, the object is only drawn there. For example, have you ever thought that you saw something, and then looked again and it wasn't there? That probably was a millisecond neuron writing to your eye neurons. Another classic is drawing (mapping) the image of somebody you know over the face of somebody you don't know for a millisecond. I have no idea why those with neuron writers do this. I have experienced this numerous times, and even my Mom recollected this phenomenon which happened to her while in college - in the 1950s.

The "Thought-Screen"

Just like there is an "eye-screen" that shows what a brain is looking at, so there is a "thought-screen" that shows any image a brain is thinking of. It seems likely that, most of the time, a thought screen is simply black, which indicates that the owner of the brain is not visualizing anything in their mind at that instant. Think of an image of some food that you like to eat. That image that you just thought of appears on your thought screen and can be captured and sent to other people as an image (.jpg) file. For many people, it may be that only occasionally and briefly, an image will flash on their thought-screen which shows an image that they are thinking of or remembering. For example, when an image of food appears, it most often means that the person is thinking about what they want to eat. An image of a place the eyes saw once may appear, for example, the image of the view in a grocery store aisle, which may indicate that they are deciding if they may go to that location sometime in the future. An image of a person shaking hands with somebody they see means that they are thinking "should I or will I shake this person's hand?". It may be that D2B consumers have many more images on their thought-screens, knowing that such images can be seen, and have trained themselves to draw images there. A thought-image can convey a person's instant fears or impressions – many times in a comical way- for example, a person asks another person if they want to buy a book, and the receiver of the offer has a thought image of themselves sleeping with a big "zzzzz" over their head. So the thought screen of a D2B consumer may be much more animated and brightly filled than the thought-screen of an excluded, but I don't know for sure.

One really amazing aspect of both the eye and thought screen is that even other species have an eye and thought screen. There is a distinction between just seeing something, and being able to "redraw", or "remember" the image the eye captured in some array or layer of neurons. The ability to draw or remember images on a thought-screen probably is found in all species back to the evolution of fish. Ctenophores (TΘNAFORZ[1]) and Cnidarians (NƀDAREꞁNZ) (sea pens, sea anemones, jelly fish,

---

[1] See my one-letter-equals-one-sound phonetic alphabet to hear how each letter sounds at http://tedhuntington.org/fonik.htm.

and corals) are the most primitive multi-cellular organisms to have neurons. Many jellyfish and other invertebrates like worms definitely have an eye screen, and also may have a primitive thought-screen which occasionally shows images. I think certainly many birds and mammals and probably even reptiles, amphibians and fish can produce images written on their thought screen. Which species is the most primitive that can draw to a thought-screen?

"Ears", "Hearing", "Ear-neurons", "Ear microphone"

Not only is everybody a "walking camera" because their eyes capture images, and a "walking billboard" with a screen that shows their thought-images and images externally sent or activated there, but we are also walking microphones that record sounds around us. In addition to the visual neurons of the "eye" and visual "thought" screen of the brain, there are auditory neurons that can be read from and written to. "Ear neurons": *Reading* from neurons of the ear allows people to hear what the body hears externally, for example, a song from a speaker, a bird chirping, a car passing, etc. – but all done remotely, through invisible frequencies of light particles, without having to touch the brain. *Writing* to the ear neurons can make the owner of the brain think that they hear external sounds, like a song from a speaker, a bird chirping, a car passing, etc. again, all done remotely using invisible frequencies of light particles and many tiny floating and stationary devices inside and outside of the body.

"Thought-Audio", "Thought-Sounds", "Inner-Ear"

Just like there are auditory neurons that record external sounds, so there are auditory neurons that record the internal sounds of thought. To *read* from these neurons allows a person to hear the sounds a person is thinking of. For example, think of a song. Then think about being able to record that song in your thoughts and send it to other people just like an mp3 sound file, again all done invisibly with low frequency light particles. To *write* to thought-audio neurons produces sounds in the body's thought-audio. For example, sending the sound data (like from a .wav sound file) of some song to a person's thought-audio neurons, makes them hear the song, not externally, as if the sound is being playing from a speaker, but *internally*, as if the sound is located inside their head. It is, of course, possible, to play audio of the voice of the brain's owner, which might trick a person into thinking that they themselves actually thought something that was externally written to their thought-audio. Even activating neurons that form internal thought-audio sounds might be possible, which really make a person feel that they truly created the thought-audio themselves.

The hearing and thought-sounds of the other species can definitely be read from and written to. Imagine how interesting that would be to those of us who have never heard thought-audio from a different species. For example, hearing a horse play back a song from the radio, or recalling a human voice. Which species were the first to have thought-sounds?

Other sensory neurons

Presumably, any neuron, inside anybody, can be read from and written to, and there are many complex neurons. A brain can be made to feel sad, to laugh, to cry, to remember some image, sound, smell, taste or physical sensation from long ago in the past- presumably just by making that particular neuron fire, or perhaps by making that neuron have the highest relative electric potential (voltage) of all the neurons in the cerebrum.

In addition, neuron writing is used to make you "aware" of some part of your body, that is, to make some small part of your body the focus of your current thoughts. Neuron writing can make you feel pain anywhere in your body, can make you feel like you are getting an enema, that your heart is pounding, and the feeling that you need to itch some part of your body. Beyond remote neuron writing, it seems likely that there are microscopic particle devices that can simply make a person feel an itch on their skin, without neuron writing, but just by sending light particles at the skin. But activating a neuron organelle is probably the standard and most commonly used method. One response to the constant remote molestation is to fire a mental laser back at the particle molester in your mind.

Motor neurons

Because there are neurons that are connected to muscles, which are called "motor" neurons, remote neuron writing can be used to contract any muscle of any body that has muscles – and all multicellular animals except sponges have muscles. Many people do not realize how important muscles are to live. There are many muscles that can be remotely contracted to seriously hurt or kill a person very quickly, for example simply remotely holding the lung muscles, contracting arm muscles when driving a vehicle, or leg muscles when near a steep ledge.

Figure 1. An office meeting where 4 people are included and 2 people are excluded. Included people probably have a D2B window of each other person in the meeting. Notice how those who are not getting D2B look "sterile", and disconnected. See d2bw.org for a closer look at the above scene in a video.

So, direct-to-brain consumers probably have a number of windows in front of their eyes. Probably people mostly have a window constantly open for each person that they care about: their mate, family, friends, coworkers, and neighbors. Each window probably shows one circle (or oval, or rectangle) with a live image of the person from the front (the person's face), another circle with their eye image, and another circle with their thought-screen, in addition to being able to see the D2B windows that the person may be looking at. Perhaps they only just get updates of relevant videos without having a constant live image.

D2B consumers don't need to look at you to see you, but you need to look at them to see them. It seems likely that when a D2B consumer walks by anybody, a D2B window must pop open with: 1) a live image from the front of the person walking by (their face), 2) a window with what that person's eye's see, 3) a window with their thought screen, and their thought-audio starts to play in the thought-audio of the D2B consumer. So an excluded has to look to see what a person looks like, while a D2B consumer can stealthily and sneakily watch a person around them for a long time without having to look at them. Not only does the D2B consumer get a live image from dust sized devices around and inside the excluded person, but probably gets a window with "highlights" from your life (your most memorable moments), and/or a quick profile about your beliefs, whether you are married, have kids, etc. We all simply want to know about those around us to protect ourselves, to look for friends and mates, but also to learn more and simply to "people watch". You can see how, obviously, this natural curiosity, fear of the unknown, and desire to know the actual truth is what drove and still drives the creation of microscopic devices that fly around and send back images from inside houses and of thoughts. There may be a quick set of ratings about the person, like how many people have rated their physical and intellectual qualities. One important note is that there may be some kind of "privacy" custom where D2B consumers don't get to see other D2B consumers in their houses or their thoughts, but you can be sure that no such custom or courtesy exist for those of us who are excluded; even the worst D2B consumers can probably see inside the house and head of any excluded person they want to. D2B consumers know that they can just look at our thought screens and thought audio and inside our houses without paying us a cent, whether we want them to or not. This includes even writing to excluded minds, because unlike a D2B consumer, the payer and the writer taking the money to neuron write know an excluded cannot see them or put up a defense in any way – it's like beating up and taking candy from a baby or defenseless child. And then there are always a million flimsy excuses as to why this kind of barbaric segregation and inequality is ok.

Think of the worst things you have ever done, or the most sexual, and/or most embarrassing or most violent things you have ever done. Probably they were all remotely neuron written suggestions, as crazy as that sounds. Those are probably the first and most popular videos of you that the D2B consumers see. We all have them, so you are definitely not alone – some are much more violent or sexually embarrassing than others. You went out in the nude, you lied, you stole, you assaulted somebody, you wore opposite gender clothes, you did something sexually inappropriate, unusual, or embarrassing, etc. D2B consumers have far less of them, because, of course, they know to reject remotely written suggestive images and sounds, and that anything they do will be used as a weapon against them for the rest of their life, and may greatly restrict their monetary and mating possibilities. The D2B owners send constant sexual, suicidal, violent, unhelpful suggestions in many different forms, very fast thought-images, thought-sounds, etc., and excluded people have no idea that somebody is sending them some terrible suggestion, they simply repeat it.

Because people can hear your thought-audio, thinking thought-audio is like talking. It's very tough to remember this fact, in particular when you don't hear the thought-audio of other people. D2B consumers may have learned to stop all thought-audio, for example during meetings (see fig 1), because producing thought-audio during a meeting is like rudely talking out of turn. But excluded don't know about this social practice, and so listening to the thought-audio of meetings and group gatherings must be kind of funny, because one or two voices-the thought-audio voices of the excluded people-continue on while all the thought audio of the D2B consumers is quiet. But also it must feel disgusting, to leave other humans in such a 1200s kind of state.

Walking down a street, passing people, you have to realize that a little window pops up of you, your eye screen and your thought screen, and your thought-audio starts playing in the D2B consumer's thought-audio, which they can control the volume of, and completely mute if wanted. But probably, like talking, most people probably like to hear all the thought-audio around them. So walking by people on a street, you have to realize that your thought-audio is like you are talking out loud – the D2B consumers hear it all. They also know what you are looking at, in particular, if you are looking at their butt, breasts, etc. because they recognize the picture of their butt in the window showing what your eyes see.

If we see an image in our mind, for example the image of a funny face, we decide quickly if we want to echo, or act-out, that is, move our muscles to imitate that same funny face. Neuron writing is many times just to plant a suggestion, and to make the context of thought some particular topic. For example, they write an image or a smell of a pizza for 1 millisecond, just long enough so that you notice, and your mind then decides if you want to eat pizza, making a yes, no, or no decision. Neuron writing is the same thing as seeing an ad, but these ads are sent directly to your brain without you seeing the sender, without the receiver knowing even that such technology exists, and so they are very persuasive. Once a person realizes that humans are sending images, sounds, and smells they can step back and resist an instant acceptance – or at least have some skepticism.

Is that song in your head you just happen to find yourself humming sent there by people with bad intentions? Probably. One surprising truth of remote neuron writing on excluded people is that many popular songs may have been partially secretly remotely written by violent wealthy people who buy remote neuron writing to influence a large audience. There was a funny skit on MTV where Randy of the Redwoods is reading a newspaper story out loud about how a killer is coming after him, while a man approaches him and appears to be following the story line Randy is reading, and so then Randy ad libs by changing the story to read that the supposed killer then goes far away, prompting the approaching man to move away. This is a good analogy for the way neuron writing is used on excluded people to try and get them killed. One example is a lyric in John

Lennon's "Out The Blue" which is "...you came to me, and blew away life's misery...", could this have been remotely written onto Lennon's brain without his knowledge, that is, without Lennon realizing that such a message might be misinterpreted to mean "blew away" with a gun? I myself hear many songs in my head that, me repeating and whistling out loud, would not be to my best advantage – songs by notorious homosexual people, for example. I am completely a supporter of all bisexual people, but this kind of neuron writing is used to try to falsely label or influence excluded- to make an excluded person appear to be homosexual. Terrible D2B owners and consumers constantly try to explain the apparent lack of success excluded people have finding mates as due to homosexuality. But in reality, the lack of success at finding mates is obviously because of the neuron apartheid- excluded people simply cannot see potential mates and their thoughts. Much of the remote neuron writing is designed to divide like-minded decent people by focusing on non-violent issues like "gay?", "perv?", and "insane?" instead of "violent?". The argument given by the neuron consumer buyers and neuron owners for such suicidal, homicidal and unpleasant neuron rape writings of humans who have never even heard of remote neuron writing is that it is simply "free speech".

Classic remote neuron writings

There must be some often used traditional remote neuron writings that are centuries old and have a long, but publicly secret, history.

<u>Motor neurons</u>

Lung muscles

Of course, a person's lung can be held so that they are suffocated – and because the particle frequencies are invisible, and the remote murderer cannot be seen, there is nothing any coroner can say other than- "here is a perfectly healthy person that just died, without any evidence of a cause of death". Holding the lung muscles may be one of the most popular forms of remote murder, and probably more than 1 million people have been killed by somebody, or by a group of people remotely keeping the lung muscle of the victim contracted or uncontracted until death by suffocation. Many times a lung muscle may be remotely and invisibly held when a person has suffered some kind of damage, or has cancer and is in a hospital, or uses recreational or prescription drugs, because there is much less suspicion when they suddenly die. It may sound crazy, but illness, cancer, and drug addiction are helpful excuses to explain why a person mysteriously and suddenly dies. Possibly among those killed by this method of suffocation are, not only my Mom, but many famous popular liberal people like: Michael Jackson, Elvis Presley, Jim Morrison, Jimi Hendrix, Janis Joplin, John Belushi, and Marilyn Monroe.

Arm muscles

When a person is driving, the remote neuron writer contracts an arm muscle to make the victim drive into an accident. Could this be what happened with

Teddy Kennedy, and a person speaking out about 9/11, Danny Jowenko?

Another classic motor neuron writing is moving your arm and/or hand to make you cut yourself. When you are cutting a vegetable, remote neuron writing can move your hand slightly to cut your finger, and the same for peeling a vegetable, or to burn your hand on a hot rack in an oven, or to get stabbed by a knife in a dishwasher, etc. I have even had my hand cut with my own fingernail. Computers analyze sharp objects, and can make use of them to hurt you in their 3D rendered model of you and your surroundings. One piece of advice is: don't hold onto sharp objects like knives for a long time, carefully and consciously make each cut, and when using sharp objects, like drills and machine tools, get in and get out, because advanced computer programs can easily remotely and involuntarily move one of your muscles a few millimeters to cut your body.

Making you cut yourself during shaving is common; by simply applying a little more muscle contraction than is required, or moving your arm just a millimeter too far to cut your nose, etc. There is really no way of protecting yourself, because the remote particle technology is obviously more powerful than we are, but you may be able to reduce the remote cutting by taking extra precaution when using and working around sharp objects.

Leg muscles

A quick muscle contraction can send a person off a roof, bridge, cliff, out the window of a tall building, into another car, etc. We can be optimistic that self-driving cars will remove much of these remotely caused accidents, but even remote transistor writing will probably be a problem. The best answer is to identify and stop the people behind all the violence-causing remote writing.

Finger muscles

Making a finger twitch is popular. One use is to deprive a person of sleep. Just when the victim falls asleep, their finger is made to move, and this wakes them up. Once awake they cannot go back to sleep. Sometimes a right middle or index finger will be twitched to indicate that the victim is not "right-wing". A finger might be twitched to make an excluded person think that they have a health problem (like carpel tunnel syndrome), or feel less confident because they mistakenly believe that they are getting older and their fingers or hand is shaking like an older person.

Sometimes tiny cuts can appear on your body – mostly on your fingers – and then your right middle or index finger. Perhaps these cuts may be done with particle devices that do not communicate with neuron organelle devices, but probably the neuron organelle devices can be made to move around or cause similar damage within the body.

Eyelid muscles

The tiny muscles that control the eyelid are sometimes made to twitch, and this is supposed to imply that you have a mental disorder.

Eye muscles

Moving your eye muscles is one of the trickiest muscle moves, because it's hard to believe that even where you look, where you point your "camera" (your eyes), can be remotely controlled. Eyes are moved to make you look at something: a knife, a butt, an ad, a license plate, etc. Sometimes there might be a two part molestation, like your eyes are made to look at a knife as you walk by it, and around the same time, an image of a friend is flashed on your thought-screen. You can easily see the quality of the people that run remote neuron writing.

Vocal muscles:

There is also the so-called "Freudian slip", which is when your vocal motor neurons are written on to make a verbal mistake; for example to say something that was not what you intended to say, but that was somewhat comical, although usually from a low-brow perspective. Probably 99% of so-called "Freudian slips" are remotely written and are evidence that very fast computers not only understand the perspective of the person paying for the neuron writing, but also what makes people laugh. Because the computers are faster than the humans, perhaps a person just buys some verbal neuron writing without knowing what specifically will be done, and the computer just works with what the victim is saying at the time.

Sensory neurons

Touch neurons

Itches: For example, you have to scratch your face and in the process you make a right or left arm salute. The "left" "right" analogy is very popular to the wealthy who constantly do neuron molestation, probably because a small group of wealthy violent criminals wants to hide behind a large group that identifies with being "right", but also to focus poor people against each other with constant "right versus left" fighting instead of at them. "Right vs. left" is a major theme for the evil side of the neuron. Like the "Reich" (right) of the Nazi's, perhaps one of the biggest convincing divider of people is this concept of "right" and "left". The murderers, assaulters, remote particle molesters, and liars all band together under a banner of "right" and all those that refuse to obey or ever disagree are, of course, the "left". "Blue versus Red" has also been a popular method to divide people, but "left and right" is apparently more popular for the neuron. Directions like "middle", "up" and "down", "forward" and "backward" are not as popular. As an analogy with the human body, a more logical view is that both left and right sides should be equally utilized, exercised, trained, and developed- anything else results in a body that is not symmetrical. Beyond that, how violent and dishonest a person is, is a much more useful, clear, and important way to distinguish between people.

Ear neurons

Remotely writing to ear neurons is perhaps the number one most popular form of remote neuron writing. Many times the audio sent to a person's ear is some kind of verbal abuse. Many times a gruff voice in your head may say "perv!", "vert!", or "nut!". Other times a female voice may say similar words, probably because the 90% male owners of neuron writing want to somehow shift blame and anger to the mostly powerless women of Earth. A bizarre part of the D2B is how they all secretly watch people's thought screen while the unknowing excluded people are showering, but somehow the excluded are the "perverts". Writing thought audio like this may be to reinforce the myth that somehow those who own and receive D2B are sexually and mentally normal, but that those excluded are excluded because they can't control their behavior and their sexual desires, or have unusual behavior and unusual sexual desires. Also the goal is to change the focus to an antipleasure and sanity context and away from exposing the murderers, going public with D2B, and other more important and honest topics. The worst part is that the hypocrites that abuse excluded people, send all kinds of sexual suggestions – to go nude in public, to put a banana in your butt, to put a door knob in your butt, to wear the opposite gender clothes, to sexually touch people- in particular people under the age of 18, etc. Then the excluded bites on the suggestion, and may get in serious trouble, lose many friends, their job, and be viewed as a fool, but all those distant neuron writers are never even seen, let alone blamed or punished as being probably the main reason that any kind of unusual behavior occurred in the first place.

Other classic thought audio writings are "shutup!" and "rat!". Ironically, the neuron writers never shut up with their constant thought-audio molestations and big lies, and while you can't see one image of the offending writers, they have unlimited access and freely share (or "rat") all your information including images of you in your room, and your every thought-image and sound.

Some of the verbal abuse in an excluded ear is used to reinforce a myth that the excluded is dangerous, like a woman's voice saying "leave me alone!". But, of course, any thought of leaving the excluded person alone and not writing to their ear neurons is not of concern. Sometimes the audio "leave me alone" is flashed with an image to associate the audio with a person. An excluded might become angry at such an accusation and wrongly blame the actual owner of the voice or person in the image. It seems almost certain that the voices and images of people sent with D2B are most often "identify theft" and impersonation. Synthesizing a voice and sending audio samples is a simple thing to do, and given all the violence done by the neuron owners (9/11, the Kennedy murders, all the galvanizations, etc.), impersonating other people by sending sounds of their voice and images of them to people's thought-audio and thought-screen is minor in comparison. Probably many excluded people instantly believe that their friend really did say the terrible audio in their mind, even though it never happened; the audio and images were synthesized and remotely written to their thought-audio and thought-screen.

The technology has evolved over many years, so sounds like the creak of a chair, or door can be "shaped" in the mind to sound like a voice. A distant voice can convey some paid-for remote neuron message of hate. The indistinguishable sound of people talking in the distance can be shaped or similar sounds overlaid to make a person recognize some audio phrase that is never actually said.

Eye neurons

On excluded people, eye neurons are more rarely written on, and when they are written on, it is usually extremely fast. But when a person is sleeping, there may be a lot of remote writing to the eye neurons. In fact, probably most of the remote writing to eye (and thought-screen) neurons is done while an excluded person is sleeping, or when they are just coming out of deep sleep. The science fiction writer Hugo Gernsbach talked about people of the future remotely learning while sleeping by using remote neuron writing way back in 1911, and there may be very wonderful dream movies that people can buy to see while they sleep (and interactively be a part of – as if you are in the movie), but in my experience most of the movies sent to dreams are terrible. I've had dreams where my friend had their head cut off, where my Mom jumped off a balcony to the pavement far below and died, where a friend was chopped with a hatchet, where I was chased, where I was shot at by a person with a handgun, in one my Mom was shooting a gun at me, where I was trapped in a hall that kept getting smaller and smaller, where a door opened when I was alone and a person suddenly attacked me from the darkness behind the opened door, where many insects like spiders were chasing me and crawling on me – many times the big insects are stuck on the dreamer's body- like on your arm and you blow and swat at them (while still dreaming), but they don't get pushed off– those are just the ones that come to mind. Because the dreams are in front of your eyes, it's exactly like these terrible events are really happening around you, and I'm sure can cause a heart attack or stroke just from shock and surprise of what people experience in these scary and violent neuron written on dreams. I often awake, from bad dreams, and knowing that they are sent there rapidly fire back at the senders in my mind, but it's probably useless. Probably the evil half of the Neuron sells these videos to like-minded consumers who then laugh to see their enemies tortured in their dreams. You may even recognize the dreams I describe above, because they probably get replayed on the eyes and ears of many poor victims. On the opposite side, very rarely, there have been a few sexual dreams that were nice, like where I got to kiss a pretty woman or got oral sex, and then there was a dream where there was a complete musical and the chorus of the song that many of the people sang and danced to, which I still remember even now, was "when you feel some motion!". But I never get any movies where we go to the other planets, the other moons, the other stars, or meet living objects of other planets – just most often the lowest brow, most terrible dreams. It's probably no wonder that we receive stupid and scary neuron written dreams, when so many people bizarrely embrace "horror" and other violent movies, and the idiocy of religions. All this time we could be seeing the thought-screens and dreams of people and the other species, but the wicked D2B owners have kept it all for themselves.

When, on the extremely rare occasion that an image is sent to an excluded thought-screen, it is almost always an unwanted image, for example an image of a person who you don't like or don't get along with, or even simply don't want to be reminded of.

Other neuron written images you might receive:

Images of food. Probably obesity and being excluded are closely linked, because people constantly send images of (and impulses for) food – but so quickly that a person who is not aware of D2B doesn't realize that the food images (and impulses) are being sent to their brain remotely by another human. Then, of course, a person decides whether to eat the food in the image. The chances of an excluded person blindly following the suggestion are probably high. Many restaurant chains and food companies may send images of their products because they know from experience that they get more business by planting the suggestion of their food item on the thought-screen of excluded people. Those companies who don't neuron write ads may lose money or fail altogether. Making an excluded person fat also lowers the popularity, and therefore the influence of the excluded person. But more sinisterly, remotely neuron written food suggestions to make excluded people get fatter are used not only to increase an excluded person's risk of health problems, but also to lower their chance of getting sex, and therefore of reproducing, and so are used by wealthy people to try to lessen the influence and ultimately to exterminate their enemies (political, religious, philosophical, etc. opposites) among those who are excluded.

Images to make you do something sexually inappropriate. When the goal is to lower the popularity and chance of reproduction of a smart, honest, talented, educated, good-looking and/or heroic excluded person, making them bite on inappropriate sexual suggestions tops the list of neuron writes. Planting an image of an underage friend while the excluded masturbates so the excluded person then includes images of the underage friend into their masturbation thought-images is probably one classic technique. Thought-images that people have during masturbation (and sex) probably generate a lot of money for AT&T and the neuron owners. Excluded people tend to be more open minded about sexuality, in particular since they don't realize that many people are watching them and their thoughts. Much of the key to getting excluded to do sexually inappropriate things, is that the excluded simply has no idea whatsoever that they are always on "candid camera". This idea is unheard of; they can't possibly realize that microscopic cameras and neuron reading and writing devices are a very common part of life on Earth, no matter where they are.

Images to make you steal. Another way to lower the popularity of popular excluded people is to get them to steal. It's amazing how easy it is to get a decent excluded person to steal when they don't think that anybody can see them.

Other images. I have seen images of a tattoo on my body for a millisecond when I look in the mirror, and thought "how would I look with a tattoo?" and "Would I ever get a tattoo?". It prompts the question: "Get a tattoo Y/N?" in the thought-audio of excluded people. I have seen piercing jewelry (recently one in my navel). Probably at least 20% of excluded can be made to get a tattoo or piercing using this method. Although some people find tattoo's attractive, creative, and a turn on, probably for most people a tattoo (like obesity, violence, and other acted-on D2B suggestions) can significantly lower a person's reproductive and economic value. The senders of the D2B image know this (but most excluded don't), and this method of "lowering the value" of political enemies (usually those for democracy and majority rule), and mostly excluded people, is an industry that many excluded are the victims of.

As an excluded who has never really seen a thought or eye screen – it's tough to see how people act and think, but for those people who have been watching eye and thought screens for centuries, human behavior must appear to be very simple. So I can only kind of give you a basic view – videos that show thought-screens would much more quickly and effectively show you how people think and how they are tricked with remotely sent images and sounds.

Memory neurons

One very subtle form of remote neuron writing is to activate memories that then relate the message the buyer of the D2B writing wants, for example, to activate neurons of unpleasant memories – or memories of events where the evil side "won" – like a successful shill from a person who normally resists the money to shill. Remotely activating an earlier memory may be less expensive then writing a new thought-audio or image, but also, may feel less artificial and less externally produced to the victim since the memory is one they have felt or seen before.

The neuron "organelle"

For a long time I thought that remote neuron reading and writing was done just with light particles in x-ray frequencies directly to the neuron, and maybe that is true, but it seems likely, as crazy as this sounds, that micrometer and/or nanometer scale devices, like some kind of RFID devices, which are very small – smaller even than dust, may function as the first human-made cell organelles (like mitochondria or plastids but made by humans). Another theory is that particle devices, probably microscopic, send light particles of invisible frequencies right into the eyes and ears to draw pictures and create artificial sounds without any intracellular devices. Such external devices could also do remote neuron writing directly on the neurons of the motor and pain (nociceptor) neurons by penetrating the outer skin (which of course we all know high frequencies of light particles can do). I doubt such a device could do remote neuron reading very well. Possibly, they could see eyes and hear thought from the sub-visible frequencies of light emitted from the brain, but since there are many layers of neurons emitting light particles, it might be difficult to isolate particles emitting from just neurons responsible for smell, for example. Maybe this kind of technology was developed initially, before the "ah much better!" moment when a neuron intracellular electronic device that assists in remotely communicating with specific individual neurons for reading and writing was invented. To my knowledge, nobody has ever published a method showing how to remotely activate neurons using only light particles, but there is very little published on those kinds of experiments. Without an intracellular device, I think it would be very difficult to read from or write to individual neurons, but such technology might be accurate enough to remotely make a muscle contract, cause sounds to be heard, to make a person feel an itch or smell a smell, and to make an image appear in the eyes.

It seems likely that to remotely read and write to individual neurons, in particular those deep inside the brain (like activating a past memory), there needs to be intracellular devices. These tiny dust-sized devices can enter a body through the lung, enter the blood vessels, and eventually enter various cells. These devices might use a network (MAC) address, be remotely steered and moved with tiny motors, transmit the electric potential of the neuron or any other cell that they are in, and be able to change that electric potential – to make a neuron fire. A neuron is a lot like a transistor, or electronic switch that is either on or off. One of the best pieces of evidence for this theory is the scene from the 1967 movie "The President's Analyst" where the phone company representative explains how a chip enters the body and is used to make phone calls using only thought. Since remote neuron writing using only light particles has, so far, never been proven publicly, it may be that some kind of device needs to be in a neuron to make it fire. In addition, having an electronic "cell organelle" device with a numerical address, removes the problem of directing a signal to some precise neuron location; instead a stream of light particles in an invisible frequency, that contain an encoded address, can be sent in the general direction of a person's head, and only the specific device with the correct address will respond to the signal, firing the specific neuron, or sending back the neuron's value. Another strong piece of evidence in support of the theory of very small chips functioning as cell organelles is that this is the basis of some of Galvani's famous 1780s experiments. Many people don't even know that Luigi Galvani did remote neuron writing by 1791 using a spark machine and holding a scalpel directly on a frog nerve. Imagine the process of cranking the spark generator as still the same, but replace the scalpel with a much smaller piece of metal. The metal amplifies the light particles moving from the spark, raising the electric potential of the nerve cell causing it to fire. What is shocking is that, not until November 5, 1959, with the remote control pacemaker, was there one published report, to my knowledge, of this simple extension or optimization of Galvani's spark and scalpel experiment of making a neuron fire remotely, but with modern technology or even a much smaller cell-sized piece of metal. This is not to say that remote neuron reading and/or writing cannot be done without such a device, certainly remote neuron reading can (as Kamitani, et al showed in 2008). Perhaps remote neuron reading and writing without an intracellular "organelle" device was how RNRAW was first uncovered way back in 1200 or whenever it was first discovered. The history is completely secret, but clearly there must have been transitions like: the first "micrometer", and first "nanometer" scale device, etc.

It's hard to believe that there could be microscopic communication devices in our bodies. But for comparison, look in sunlight, for example in your car, or in the light under a lamp against a dark background, you should see thousands and thousands of tiny pieces of dust and fibers. Those pieces of dust, that we normally don't see, are perhaps 1 mm in size. By today's electronic standards, 1 mm in size is actually very large. The Hitachi "uChip" RFID chip is micrometers in size. So looking at those thousands of pieces of dust, it is easy to imagine that we inhale dust all the time. And that those tiny dust fibers enter our lung and some of the smaller objects then enter our blood stream. Even some microscopic diatoms float in the air and enter our lungs. These tiny devices can easily be powered by ambient light particles, even of lower frequencies like radio and heat, for example that exist inside the body, or even from the natural chemistry of molecules inside a body.

There probably has evolved a complex system where the chips inside cells (which number, similarly to the number of neurons in many animals, in the billions for each organism) probably have sophisticated schemes of changing their MAC address. The chips may regularly change their address according to some internal algorithm which is only known to the manufacturer, and probably many have the ability to be remotely programmed to change their MAC address. There must be massive systems of tracking and storing the many millions of changing MAC addresses of the countless intracellular nanometer scale devices. Probably a stale-mate has occurred because both sides can easily write to tiny devices that cause harm or annoyance to the other side, and their battles involve contests of how much remote damage and annoyance are they willing to inflict and simultaneously endure before stopping.

Couldn't we find such a device if they are in every neuron?

Possibly somebody could, in particular somebody who dissects animals that has access to an electron microscope, but probably most of the people in those fields are either D2B consumers and sworn to secrecy, or excluded and don't have any idea that such devices could be there. But even if there was an excluded that wanted to, there is one really terrible but interesting possibility: that with remote neuron reading and writing, any object, in particular small objects can be made "invisible" to the owner of the brain simply by overwriting the signal to the neurons. This may be a basic component of remote neuron reading and writing – to basically "erase" any floating remote neuron device that reflects light into the eye of a person. The neuron readers can analyze the image captured by the human eyes in milliseconds, long before the human brain can, and so it can then cover over any device – as crazy as that sounds. It's similar to the classic question of – is what we are experiencing actually there or is it all being remotely written to our neurons, and we are actually in some extremely different place? Alternatively, could we be made to live in a permanent alternate reality simply by remote neuron writing?

Remote particle molesting and assaulting

There are apparently "career" remote particle molesters who do nothing but pay and get paid money to remotely move people's muscles and write annoying neuron writings – in particular on to the brains and bodies of excluded people. These scummy D2B writers use remote neuron writing to make people itch, to deprive them of sleep, to make them stub a toe, to spill a drink, to drop some object (most usually soap), to make a gesture (most often a right or left "salute"), to cut them, and similar such annoying remote neuron writing.

The arrogance of many D2BW owners and consumers is unbelievable. Not only do they get to see inside houses, but they also get to routinely see and hear everybody's thoughts- they have unimaginable advantages over the excluded people of Earth- but that isn't enough for them- they have to remotely molest, assault and murder too!

It's unbelievable that such idiotic, annoying and petty people could be in such high positions of power and so involved with remote neuron writing – the "neuron" is absolutely upside down without any doubt and constant particle remote molestation is one of the best pieces of evidence of that. The tiny good that does come out of it is that a very few excluded people may start to wonder about remote particle devices when they feel an itch, or one of their muscles is moved. Sadly, most people probably visit a doctor and think that they have some kind of disorder like carpel tunnel syndrome. As I said in the introduction you almost certainly are a victim of remote neuron writing on many occasions and did not even realize it.

Speaking for myself, and no doubt, many other people, I am constantly being remotely molested (made to itch my face, sent some "prank call" thought audio put-down, my muscle made to stub my toe, made to say something wrong or to forget something, getting accidental cuts, etc.). One possibility is that some people are remotely molested unless they remotely molest somebody else (for example an excluded person), and so you can see, how, in the tough, survival-of-the-fittest environment we all live in, how there may be a constant remote particle "tug-of-war", "shoving match", "kill or be killed" kind of struggle going on. It may be that somewhere there is a much more violent and painful remote particle tug of war going on, which is reduced to just some smaller remote molestation by the time it reaches us.

Some remote molestation is very subtle. For example, when a person is going to speak publicly, remote neuron writing may be used to make their heart muscle pump strongly, or to move their vocal muscles to quiver a little, which may cause a loss of confidence. The best response is the usual "fire back in your mind". Then you find that, many times, as if by magic, your heart stops pounding, and the nervous feeling suddenly goes away. Many times if you receive a positive, supportive, thought-audio, you are instantly hit with terrible remote neuron writing- a common example is your thought-audio voice commanding the positive thought-audio senders to kill themselves, or a put-down to them, and other similar terrible neuron writings. I usually immediately fire back on the remote molesters in my mind, and often try to counter with positive and supportive thought-audio to the positive thought-audio senders.

The voice and images in our thoughts are our most intimate, influential, and trusted advisor. For that reason if a technology was invented that can remotely and invisibly write sounds and images to a person's thoughts, you can see how extremely irresponsible keeping it a secret would be. In particular, people are starting to realize that, not only is that voice (and that image) in our head not the intimate, influential, most-trusted advisor we thought it was, but that most of the sounds and images in our thoughts are being sent there by our enemies- people with a violent, dishonest, and criminal history. Those who own, control, and participate in remote neuron reading and writing should show even the tiniest bit of decency by at least *showing* and *telling* the world about RNRAW in the interest of protecting all those inhabitants of Earth that are being unknowingly abused by it.

Other subtle remote writings, are the negative feeling or impression you may instantly get about your friend or another person you are talking with or walking by. Remote neuron writing is used to create and amplify negatives, criticisms, and flaws, and to avoid positive thoughts, like compliments. Many times that voice in our head, or even a very subtle negative feeling in our mind, suggests that our friend isn't cool, or is weird, is unfashionable, is not very smart, has ugly physical features, is mean, has flaws, etc. Some of these "instant reaction" thoughts may be internally generated and natural, but it seems possible that many might be remotely sent there by evil violent people on the other side, as strange as that sounds, that are trying to defeat their enemy which includes you and your friends. Probably a lot of the "voice in our head" and the feeling in our mind is sent there by violent criminals and is meant to mislead us and destroy our friendships. Sadly, there isn't really a lot we can do, other than to try to remember to ignore and counter as much of it as possible. So I find myself constantly reminding myself in thought-audio "ok maybe this person is not great, but violent people are much worse".

If you find yourself focusing criticism on allies (people with similar views as you have), for example for some trivial, non-violent difference, and not on clear enemies (those with very different views than you have: for example people who support the 9/11, Kennedy, Lennon, King jr. killings and lies), then probably it's the result of remote neuron writing. Many times there is some song in my mind designed to amplify some trivial difference among the anti-violent, and I try to quickly replace it with one of my self-made song melodies and lyrics, most often "9/11 was an inside job y'all", "tell the truth about the neuron writing", "stop violence, teach science", "journey to centauri", "sex ain't the big-azz problem, it's violence is the big-azz problem", "let the people get to see", "full democracy", "lock up the violent yeah, free the non-violent yeah", etc.

The one truth that seems clear about remote molestation and assault, is that it is a very good **opportunity** to fire back, in your mind, at some scummy person in the act of attacking you. Probably so few excluded people ever think about firing back, because they don't realize that some person is remotely writing on their neurons. In my experience, firing back in my mind helps. Not only does firing back erase the unnatural feeling the remote neuron writer gives you, but you have that wonderful feeling of firing back and really punishing some scum who is violating you and really deserves it. Any second, you have a bad thought, any second you feel an unnatural fear, any second you receive a suggestion to drive into traffic, or off the road, - any bad feeling or pain at all- it's an opportunity to fire back and really possibly punish some bad person who absolutely cares nothing about you, and is in the act of trying to kill or bother you, and deserves to be punished.

In some sense, firing back at people and machines remotely molesting us is like cleaning, because we can clean and everything looks nice for a few days, but we simply can't stop the unending accumulation of dust, and so we need to constantly be performing maintenance in the form of firing back when people remotely molest us. Finally, when you are too weak to fire back in your mind, the remote machines and people recognize that, and perhaps finish the job.

Remote particle molestation is probably a lot like feeding pigeons. A wealthy molester pays for a computer program to constantly send pop-up money D2B windows into the eyes of poor people with offers to "authorize" various remote molestations of their enemies, and there is a massive group of people who do nothing else but get money from "molestation money windows" (like the "shill money windows" many of us often hear the effects of) all day and night. It's not entirely clear, if those poor people doing the remote molestation are corrupted security people in the government, in the phone company, or in some other group. It seems likely that there are other "money" windows too; some might be for a pain, an assault, even a murder, and certainly many for pleasure-related services. It may be that these poor people receive some kind of fire back or punishment for their involvement in remote molestation- but the wealthy funders probably escape any serious punishment.

There is a good analogy between remote particle rape and a rape done at close distance, in that somebody is putting something in your body against your objections, and just like a rape done directly in person, as a victim, many times you naturally try to strike back during the rape to kill, hurt, mark, or somehow scar the person raping you. But with remote particle rape, for excluded people, striking back in self-defense can only be done by the victim through thought, if even that.

Sometimes there will be a series of three or four molestations, like a bad reminder, you fire back, then the same reminder a second time, you fire back again, and then a third reminder- like a machine is doing the writing. It may be that many people reject the first suggestion, but eventually succumb to a second or third writing. Some of the suggestions are very subtle, for example, suggestions that simply trigger ideas, like suicide, or doing violence, without any actual images or sounds appearing. It is gruesome to say, but there is an interesting aspect that works to the advantage of the remote molesters, and that is that most people do not want to talk openly about thoughts of suicide or violence (or sexuality) that are remotely written onto their mind, because people might think that they are thinking about committing suicide or doing violence and panic, and so this significantly helps the remote neuron molesters to stay free and unseen. Just to add on this point, that I will never kill myself or do first strike violence against other people, I love life and have a very strong mind that is very familiar with remote neuron writing, so if ever there is some news story that I ended my own life or did first strike violence you can be sure that it was remote murder or assault. You can see how people need to accept a new paradigm of "the thoughts in your head, are probably, most of the time, written there by other people"- and then currently, by people who are some of the worst violent criminals ever to live on the Earth.

You might think that it is impossible to turn off the write (and read) permission of a person who is remotely molesting, but yet, we have *never* had write (or even read!) permission of D2B, so it certainly is possible to turn off remote write and read permission entirely, but it's apparently not done for those who are doing violence or molestation remotely.

One positive result of the constant remote molestation is that, while completely unnecessary and absolute idiocy, it does make us build a stronger mind, and get better at firing back at molestations and bad suggestions with our mind.

Basic themes of remote neuron writing

There are some basic constant themes of remote neuron writing in the time we live in. One theme is to make the excluded people extinct. There is basically a "denial-of-service" (DOS) genocide of those who are denied the service of direct-to-brain windows and seeing, hearing, and sending thought images and sounds to and from brains. Not being able to communicate through thought with potential mates is probably the most effective part of the DOS genocide. Other methods are very subtle, for example to make the excluded fat by sending many food suggestions and therefore undesirable to mate with, and to make a person smoke tobacco lowering their chances of mating because less people will kiss them – in addition to the possibility of lung cancer removing the excluded. Another method is to make a person get hooked on drinking alcohol. A drunk excluded can be much more easily tricked to do something they will regret; some of the stupidest things I've done were neuron suggestions I bit on while drunk. Many of the suggestions are to stop an excluded from having the right words or confidence to get a date, or by making the voice of an unseen mate in the excluded mind make the excluded have a false sense of allegiance to the pretend mate, therefore rejecting potential mates that actually exist and are physically near the excluded.

Suicidal and violent suggestive neuron writings, are also another method to exterminate excluded, for example writing suggestions for them to jump off a steep edge, drive into traffic, drop their baby, throw their baby (you might think I am making this up, or that this is a ludicrous claim- but I don't even have a baby and I receive similar neuron writings all the time when I visualize holding my baby on my thought-screen– a woman who lives in the same county as me recently threw her baby off a parking structure – which certainly was remotely neuron written), or to use a knife or gun to do violence. Once you realize that these terrible suggestions are being sent by violent criminals, you realize just how sick and evil the neuron owners and many wealthy neuron consumers are. But those who are unaware, which is the majority of excluded people, have no idea that such evil people are writing to their neurons, and shockingly may interpret the neuron writing as being the wishes of a God, for example commanding them to do terrible acts of violence which they do because they don't want to disobey or be disrespectful to what they think is the voice or signs sent to them from a God. And the excuses given by these neuron criminals? "We done it for the right!" "It's right versus left!" "We done it for God, for Christianity, for Capitalism, for our side". There is always some label placed on the victim to try to justify the abuse: for example that the victims are perverts (this from multiple career remote writing voyeurs and molesters with victims of all ages), are insane (coming from people who reject the theory of evolution, and believe in the extremely unlikely claims of religions), are not 100% white, are gay, are not religious, are liberals, are democrats, are commies, or simply are excluded, they will never see us, that's just the way it is, it's free speech, and so on. Like the many murders, the Kennedy killings, 9/11, 7/7, etc.: all bizarre neuron raping, assaulting, murdering, and molesting is forgiven, excused, and forgotten by most D2B owners and consumers.

Stopping the free-flow of information for excluded while simultaneously removing any clog on the free-flow of information for D2B owners and consumers is a common theme. For example, "rat" is a commonly written thought-audio (and paid-for shill too). One of the best thought-audio replies is to think "do unto others" because obviously they freely share every image of you. See my "Rebuttals" section for more classic replies from the victims. They freely "rat" with all your information- sharing every single image of your life and whereabouts with any and every direct to brain consumer- people with violence in their past, people who routinely remotely molest, uneducated people, religious fanatics, etc.- while many doctors, engineers, teachers, and self-educated people are excluded. The idea of a "rat" is absurd given the free-flow of info for the D2B consumers, but like the 19-hijackers story, true or false makes no difference whatsoever, only party line matters. In addition, much of this "ratting", for example telling people that all matter is probably made of light particles, that globular clusters are our future if successful, about walking robots with artificial muscles, and direct-to-brain windows is like "ratting" about some basic education- like teaching people that dangerous secret of "addition and subtraction", or "how to bicycle", or that secret "evolution theory" (imagine how people would leave traditional religions in droves if that precious secret truth were let out). Some of us may laugh at the thought of keeping the microscope a national secret, but yet, where are the low cost, hand-held electron microscopes by now?

Much of remote neuron writing is used to reinforce the claim that a person is "insane" or "weird". It doesn't take a lot of remote neuron writing to make a person do unusual things that might scare or make D2B consumers watching nervous about helping, befriending, hiring, or including the excluded. The thought-audio "psycho!", "cuckoo!", "loco!", etc. are often purchased and written to the neurons of victims. Much of the neuron writings are designed to try and spin the audience- to make those watching you entertain that question of "is a person a ...(psycho, pervert, etc.)?" – instead of more important thought-audio issues – like "murderers!", and "accessories to murder!', and "Turn off the D2B for the remote particle molesters and turn it on for all their victims!".

Echoing remotely written thought-sounds and mimicking thought-images

One clear aspect of remote neuron reading and writing is that many people who have never heard of such a thing, have a tough time remembering that other people can already see and hear their thoughts, and so they repeat "out loud" sounds in their mind, and "act out" or mimic images on their thought-screen. It's really unfair, because the D2B excluded don't get a constant reminder by seeing D2B windows in front of their and other people's eyes. Even I, who have had suspicions that people could hear and see me in my home, and possibly hear my thoughts, still echo thought-audio or mimic thought-images sometimes. It may even be that neurons can be activated or inhibited to stop people from remembering important things, like not needing to repeat written thought audio out loud. Sometimes a person may want to verify that they can correctly verbalize some vocal impression, or they may want to expand on some thought-image with muscular movement. But mostly, wealthy people who fund and control neuron writing have taken a really poor view of how to use it, and so, many sounds and images sent to a person's brain, are not in the best interest of the person to repeat out loud or act out. For example, many times a sound sent to the brain may be a song that may have a message or association that is not in the best interest of the person receiving the song directly to their brain (the audio might have a racist connotation, or imply that the singer is homosexual, etc.). Once a person remembers that sounds and images are being sent to their brain, which are mostly like "prank" phone calls, and not to repeat them, but simply to ignore them, then the thought-screen and thought-sounds become more like a detached screen and speaker, that you see and hear, but that are not your own thought images and sounds. Your thought screen and thought-audio are more like advertising “billboards” where mostly very crude wealthy people place “ads”, mostly designed to make the excluded person echo some sounds that will make them look bad, or images and sounds meant to mislead them, or to slant the interpretation of events in a negative way. Presumably, the wealthy molesters pay the phone company who then write these unwanted ads onto the poor excluded person’s thought-screen and thought-audio without paying the billboard owner a dime.

External or internal neuron writing?

One classic annoyance is how a sound is sent to a person's thought-audio and it's not clear for any person listening to the victim's thoughts, if the victim thought it, or if it was externally sent there. Most of the time, in this terrible era we live in, the thought-audio is from an external source, and then is terrible audio- like racist words, anti-sexual put-down words, etc. It's like the rudest and crudest people are watching you and your friends, and sending the worst insults that come to their minds onto the thought-audio of you and your friends- like "you're fat", "you're ugly", "you're gay", "you're ..." on and on...with thoughts most average people would never think of. And the planetary excuse is probably always the same- why are the neuron writers so evil and annoying? Why, it's to fight against the enemy, of course! It's classical to send terrible audio to excluded people's thought-audio, and then those listening many times mistakenly believe that the poor excluded actually has that terrible view or belief. Even the excluded themselves may accept externally written thought-audio as if it was their own thought, and adopt the terrible belief or pickup and continue the conversation externally initiated in their thought-audio. The natural reaction is always to spend a few seconds thinking something in thought audio that will reverse or negate the message of the rude audio.

Using remote neuron writing to get excluded people to do violence

Trying to make excluded people do violence is one of the worst uses of remote neuron writing, and also very common. A suggestion might be as simple as activating a memory of a gun or knife the excluded owns, or saw once. The excluded then may entertain buying or using the weapon somehow for a few seconds. Many times there are suggestions to make a person hurt themselves, for example for a person to drive into oncoming traffic. The news is littered with stories of deaths that were most likely the direct result of remote neuron writing. Just in my mind now are a woman who walked her kids into traffic around Boston, another woman who drove her kids into a lake, all of whom died terrible deaths- and certainly remote neuron writing on excluded was what happened there – and certainly remote neuron writing was not used to stop them. It seems reasonable to believe that D2B consumers pay the AT&T D2B department to see these shocking videos of violence, as seen through the excluded eyes and as captured by microcameras floating around the scene, which I understand may sound unbelievable to many D2B excluded people. These videos are sent directly to the D2B consumer's brain appearing right in front of their eyes. So there is a simple financial reality: AT&T knows that they get a lot of money for violent videos, and so of course, they create violence through remote neuron writing then shop the videos to D2B consumers. The telecoms probably even get money from both ends- wealthy D2B consumers pay them to write violent suggestions on their enemies, and then they sell the videos for money to D2B consumers. Neither the D2B funders or owners are ever punished for any of the violence because they are too powerful. They must claim that what they are doing is protected as freedom of speech – that they are not actually pulling the trigger, but simply suggesting that the excluded person might want to pull the trigger, walk their kids into traffic, or off the road into a lake, etc. Like the bizarre story of the Old Testament of how God orders Abraham to kill his son (what wisdom, and decency there is in religions, huh?!), many excluded people think that the violent images sent to their thought-screen by evil people using remote neuron writing are a message from God, and don't want to disobey God, no matter how terrible God's request is. If only they knew about remote neuron writing, they probably would not act out and reproduce the remotely neuron written-on images of violence. Certainly much of the remote neuron writing is to try to get excluded people to do violence to themselves and to other people.

Even many of the other species are routinely made to do violence, presumably for entertainment and to create videos that people pay a lot of money to see. Classic examples are making dogs bite, horses buck (like for Christopher Reeve), bears attack, sharks bite, bees and stingrays sting (like for Steve Irwin), to get snakes, alligators and crocodiles to bite, whales to drag trainers underwater (like Dawn Brancheau), elephants to stampede, to get lions and tigers to attack (like the Siegfried and Roy attack), and probably many other remote neuron writings that cause animals to attack the many poor D2B excluded and consumers, who are viewed like dirt by many D2B owners and consumers. Often insects are remotely neuron controlled to collide into humans, as hard to believe as that may be.

It seems very likely that many vehicle crashes are the result of the criminal use of particle device technology, as unusual as that may sound to an excluded person. In particular flying vehicles are easy targets. It is probably very easy to remotely disable a vital part of any airplane or helicopter using the many microscopic particle devices that not only do RNRAW but that can do remote transistor reading and writing (RTRAW) too. Planes and helicopters should have safety devices like parachutes, airbags and emergency thrusters during this era where violent, lawless, unseen criminals dominate remote particle device technology. Given the truth about 9/11/2001 and 7/7/2005, how can we be sure that plane crashes like the famous Lockerbie crash aren't used to trick the excluded public into thinking that Islamic terrorists could do such a thing, to focus their anger against Arab people, and to help justify ridiculously enormous and wasteful military spending?

Much of the violence done is actually laughed at by many of the crude D2B owners and consumers. This is clearly seen even by excluded in everyday life on YouTube videos, and in popular cartoons like "the Simpsons" and "South Park" where chopping, shooting, and blowing people up is viewed as comedy- but even nudity is strictly forbidden. Many people in the bizarre violence-loving anti-pleasure D2B crowd laugh when their enemy suddenly falls and hurts themselves as a result of the invisible remote neuron writing. Part of the laugh response is probably relief that your enemy is weakened, but it may even be that an involuntary laugh response is written onto witnesses after some act of remote violence occurs. One vivid example of remote violence is how my Mom (an atheist who did modeling when she was younger), in her 60s, was almost murdered when she was remotely made to fall down the stairs in her home. An insider hinted to me that some D2B consumers laughed at seeing this terrible fall. One of the bones around her knee protruded through the skin as a result, and she lay there losing a lot of blood and too dazed to do anything. Fortunately a D2B friend of hers, probably broke with the tradition of non-interference, and went to her house, took her to a hospital, and my Mom slowly recovered over many weeks. Probably the "Christian" remote neuron writer(s) were not even punished for this vicious life-threatening assault, and to this day I have never even seen who it was that neuron wrote on her to make her fall, or who galvanized her to death a few years after that.

D2B Consumers sometimes make gestures and answer excluded people's thought-questions

Of course, you can think questions in thought-audio to the D2B consumers, which they definitely hear. It is somewhat unpleasant for them to have to respond with some kind of muscle movement, a nod, a vocal "yeah", a movement of the eyes, or fingers, etc. – since excluded people are prevented from simply hearing any thought-audio answer. Bad D2B consumers take the many D2B money offers to mislead excluded people, but decent D2B consumers may reject the thousands of dollars and honestly answer questions asked in thought-audio by excluded people. It may be that D2B consumers get some party-line answers from a D2B service provider with more authority, so that they themselves are not labeled a "rat" by responding to the thought-audio of an excluded person.

One classic question to think is: "Do you get D2B?" – many people do, and most often they turn their head to indicate "no", which probably means a variety of things: 1) yes, they get D2B, 2) no they don't think it should be public, 3) they want nothing to do with you, 4) they at least have the courtesy to indicate to you that they receive D2B, or possibly simply that 5) they are too afraid of "Orwell", the neuron owners, and their capo-like thug employees, to do anything other than indicate with a "no". Since head movement can be (and probably often is) triggered in excluded by remote neuron writing – it leaves an excluded person without any real certainty about how many people get D2B and how many don't.

Remote neuron writing on D2B excluded is very fast

Much of the remote neuron writing that excluded people get are images and sounds that last only a millisecond, but then resonate for many seconds, because most excluded people don't realize that you need to fire back and try hard to erase them out of your mind. As an excluded you don't get long duration, easy to see, D2B windows, all you get are unpleasant flashed images and brief sounds, and then mostly on your thought-screen, and rarely on your eye screen. A good example is that you heard a voice in your thoughts, usually a rude one word put-down like "perv", etc. that didn't come from anybody around you. Another example is when you thought you saw a snake, or a spider, or somebody you recognize but then when you looked again, it isn't the same image. Much of the writing on D2B excluded occurs when they are going into or out of sleep. In addition, a lot of remote motor-neuron writing (remote muscle contracting) is done to those who are excluded as I have already said.

Many times the D2B payers send an image, sound, or activate a memory for a millisecond and then must watch the excluded thought-screen and listen to the thought-audio to see what the excluded person thinks, while the excluded is ruminating on the millisecond image or sound, many times in a negative way. It's very tough at first to quickly dismiss the remote neuron millisecond writings or to quickly draw thought-images and thought-audio in opposition or to change the subject. Here's a typical scenario: Wealthy mass murderers want to rip apart their enemy, progressive, honest, educated and anti-violence people, so they send an image of a popular progressive onto the brain of an excluded progressive, and then pump negativity onto the excluded neurons- it's designed to cause chaos among the honest and fracture stop violence alliances- not just among progressives – but among any honest and stop violence people.

Many neuron writings are ads, in particular for food, smoking, alcohol, drugs, and other products, to try to make people make decisions – many times decisions that are not in their best interest. Many excluded might be recognized because they are: obese, smoke, drink alcohol, have theft, sexual, or violent offenses, etc. because they don't realize that D2B images, sounds and neuron activations are just suggestions and not to bite on them. People probably pay the neuron writers millions of dollars to write to people's thought-screen billboard, and radio-station-like thought-audio, which many D2B consumers may pay to watch. But the owner of those neurons does not receive one penny from that neuron writing rape. It is a simple idea that I think many people can agree on: that there must be consent, and that people should pay the owner of the brain to write advertisements directly on to their brain. The current system views excluded people as walking machines that can just be written on without permission, and without paying a penny to rent space on their thought-screen billboard from the millions of dollars the telecoms are paid to write ads and other unwanted messages there. Most of the direct-to-brain writing, from my own perspective as a person who is being denied D2B windows, is unlike television and radio, in that direct-to-brain writing is a service that we cannot turn off or on.

There may even be *advertising* on D2BW videos of excluded people in order for AT&T to get even more money out of the remote one-sided rape. How many D2B consumers find themselves thinking "I paid to see Ted, but all I see is Tide"? Adding advertising videos or images on the side, or before seeing excluded people in a D2B window, may make the remote gang rape feel more like television or

YouTube, and less like the gang rape of the poor excluded people that it is.

Many of these remotely written millisecond images on the thought-screen of the excluded are of people that they don't want to think about, or trigger unpleasant memories. So life for many excluded becomes based on how quickly they can forget that latest neuron write. These sounds and images sent to people's thoughts are definitely a form of molestation, neuron rape, and prank calls – purposely sending, or remotely activating a memory, image, or sound that annoys you. Some of the millisecond images and sounds an excluded may receive may actually represent a pleasant communication, although this is the extreme minority- perhaps about 1%, if even that. It's like the extremely limited D2B "service" that a D2B excluded routinely gets. One comical phrase is that the excluded person is just really "far out". Instead of an actual D2B window with a video, all the excluded gets is a millisecond image or memory triggered of some person. The excluded has no idea if that image was sent by the person in the image, or by somebody else. Instead of regular windows in front of all of our eyes, all we D2B excluded get is a millisecond image flashed in the back of our mind, or just a millisecond memory of something we saw in the past – the vast majority sent strictly just to annoy us, to waste our time, and to exploit the many people that might be watching the thought-screen of an excluded – with no thought of compensating the owner of the thought-screen-billboard. So in theory some (<1%) of these millisecond images could be actual thought phone calls from poor D2B consumers who are usually not allowed to call you. In addition, because AT&T and Verizon have a monopoly on D2B (at least in the USA), probably the price to make thought calls is much higher than it would be if D2B was democratic.

Why does the phone company, or the neuron owners, or whoever, send unwanted images and sounds to the excluded? What idiots pay the phone company to send these annoying thought images, sounds, and memories? Many times the annoying memory has to do with some victory for the bad people who send them, for example activating the memory of some unpleasant experience from your past – an argument with a loved one, a memory of something you did that was unethical, etc. Many remote molestations of excluded are probably meant to make an example out of an uppity excluded – to say to those watching: "look what we do to this excluded", you don't want to cross us or we will do that to you too. But also, many must be to show the audience watching the excluded person that the excluded really is a second-class citizen – not an equal – not heroic, to spin the opinions of those D2B consumers that watch videos of the victims, or perhaps just to make watching popular excluded people annoying to all involved, in particular to the poor excluded being neuron raped. It's all part of a massive war of evil against good. Many times the remote neuron annoying writing is meant to try and divide the honest, which form the other side or the "enemy" to the many violent criminals that have access and enough money to pay for remote neuron writing (like those who support the 9/11 controlled demolition). One example is sending you audio in the voice of your friend or loved one that puts you down or ridicules you in a shallow way. It's similar to the Popeye "Hold the Line" 1920s video. The strategy is to make two bunnies become distracted by fighting with each other so that the fox can come in and eat both. The evil of the D2B consumers and owners have this "divide and conquer" strategy down to a science, and it is very effective- in particular against the excluded public.

The vast majority of remote neuron writing that excluded people get are "identity theft" (again, just like the Popeye "Hold The Line" cartoon), and they are meant to break up or stop friendships, relationships, marriages, and families among the "enemy" of the wealthy who own and control much of remote neuron reading and writing. Some examples include sending sounds of the voice of your wife or friend saying “go away!”, “fat!”, “shut-up!”, “whore!”, “gay!”, or “I hate you!”, etc. – but the actual person never said or thought that. Other examples are sending sounds of a spouse or mate having sex with different people – including enemies, friends, coworkers, etc. - that never happened. Identity theft plays a large role in 9/11, 7/7, and many other violent acts: a group of wealthy people use military people to do the actual violence, but then buy news stories that blame excluded (in this case Arab) “patsies” for the violent crimes that the wealthy D2B people and their employees actually committed.

Much of remote neuron writing is "slanted" to the violent and dishonest criminal view

Most, perhaps 95% of remote neuron writing that is not motor neuron unwanted muscle moving, is extremely slanted toward the violent and dishonest side, because that is who owns and controls much of remote neuron reading and writing. Just like there is a "best of" collection for some music group, what most people get written to their ears is like a "worst of" collection. All we ever hear are put-downs; and other misinformation from the lowest and most far removed from factual minds. To a large extent the voice and images in our head may be a direct representation of the views and hopes of the D2BW owners since, ultimately, even beyond those evil D2B consumers who buy annoying neuron writes, they, the owners, ultimately decide and control absolutely anything and everything that gets sent to any brain directly. The views are always far from your own views in being crude, rude, always putting down the receiver of the neuron write in order to make them feel weak, ashamed, flawed, etc., but also some times to "appease" the receiver into happiness with false promises and allusions of pleasant futures that are, in reality, very doubtful. A good thing to realize when you receive thought-audio, images, or feelings, is to picture that a really evil Nazi is the only person at the controls that can write to your brain- because the truth is probably not far away from that.

When they write rotten audio on excluded brains or as "news" stories it is "free speech" and "free market", but when you want to buy ads and widely publish images that show and tell all about remote neuron reading and writing and all their violent crimes, it's "ratting", "crazy", and a violation of "national security" and "privacy".

Planetary Good versus Evil?

Is there a constant two-sided war occurring around many stars like ours? Clearly there is a spectrum, with stopping violence and telling the truth on one side, and doing first strike violence and lying on the other side, but perhaps this is too over-simplified. To survive, some matter is always going to eat or separate some other matter. Just like there may be a planetary alliance to stop violence and teach science, so there may be an opposite alliance that seeks to start violence and stop the teaching of science. Just as there are many people who are in favor of the public being shown and told the truth, so there must be an opposite group of people who are in favor of hiding the truth and lying to the public. Our species defines "good" as stopping violence, teaching science and telling the truth, and "evil" as starting violence, and lying. Using this definition, good and evil, like ying and yang, pleasure and pain, or sweet and sour, may forever be in a classic struggle that never ends.

The dominance of religions with their wildly inaccurate beliefs, and many brutal murders of history, over the minority of science is a clear theme of most if not all of recorded history. The massive and long lasting lies like the secrecy of remote neuron writing, is a clear indication that currently on our tiny planet, evil is in the dominant position. Why have the evil (those who do first strike violence and lie) been so "large and in charge" for so long? One mystery is that since they are almost all a "white-men-only" ultra-elitist club, how can they possibly maintain a majority and stay so firmly at the top for so long? Wouldn't those who are for gender and race equality in law and for integration of the many races and people form a much larger and wealthier group, because they are not so exclusive (they include women and non-white people)? For whatever reason, this apparently is simply not true yet.

Because of their great wealth, the views of the all-powerful evil at the top tend to be over-stated and dominate everybody's eyes and ears. Wealthy people who are good people avoid the risk of confronting the violent, so the evil of the wealthy people tend to dominate. Their focus is always on trivial, petty, and non-violent activities, like sex, drugs, mental instability, etc., while the views of our group, on the opposite side, tend to be extremely understated, and focus on only the most serious of the problems. For example, we are simply trying to even tell the public the truth about the centuries of remote neuron reading and writing, and about the many wrongly "solved" crimes like the controlled demolition murders of 9/11, which are not non-violent petty crimes where everybody lived and nobody suffered any pain, but instead, shocking, painful, and terrible pre-planned successful murders of thousands of innocent people.

On a positive note, we are all in a competition to see who can expose, weaken, and punish evil the most. Those who everybody recognizes as the most effective at exposing all the lies and violence of the evil people will probably be the most rewarded and respected if ever this current group of evil collapses.

"The shill", "money windows"

As a D2B excluded, it is very difficult to understand that much of what happens around you, what people say, and who is walking around you, is strictly the product of a payment of money. One way that the Internet is different from the D2B net is that in the D2B there are, apparently, many "pop-up" windows that are like "Vegas-style" offers for money. Mostly, the pop-up windows pay the D2B consumer to say something abusive to an excluded person (like "perv", or "gay", etc.). Some D2B consumers hint that this money is not like $100 or $200, but is more like $800, $1000 (a "k"), and more. I can imagine that it is a lot of money, because those who pay have a lot of money- to them, in particular, working collectively, a thousand dollars is like nothing- and because average people probably demand a lot of money to take the risk and potential abuse that might result from saying something abusive to an excluded person- after all you basically are probably going to destroy any friendship and kindness that might have existed before the shill. But the money box pays people for more than just verbal insults to be sneakily hurled at passing or stationary excluded- there are all kinds of offers- D2B is a rapid moving free market for those who are in it – similar to EBay with many people bidding up and the offer price constantly going up or down. There are not just money boxes for saying things, but also offers for sex (hands, sucks, coatings, coverings, holes, etc.), for dates, to physically go somewhere for money, muscle moves, no doubt assaults and even perhaps murders ... I can only imagine.

But just in terms of the money box for "shilling", there is a multi-billion dollar industry of misleading the excluded. Your D2B "friend" is almost always more accurately called your D2B "fiend" – but there certainly are those who refuse to say things for money. Shilling covers the entire spectrum: Those who refuse to take money to say something, those who say somewhat helpful things for money, those who say somewhat neutral things for money, those who say somewhat slanted/mildly abusive things for money, and those who say extremely slanted and/or abusive things for money. Probably a common thought for excluded people is "why are so many people so shockingly rude to me all the time?". The answer is simply because they are paid at least $1,000, and perhaps even $2-3k for each rude remark. $1k and more can go a long way for a poor person- who needs to pay rent, tuition, loans, etc. With that kind of money, the anomaly are those people who turn down the money windows to shill – they must be going for a higher and more long-term money that comes with good credit and non-abusive behavior.

But even if paid, why would people pay somebody to be so rude? For many reasons: 1) they want to protect their position of wealth by putting down popular people who oppose them, 2) to stop their violent crimes from being the focus of attention, 3) to isolate excluded people by making friendship with a D2B consumer less likely, and to solidify the segregation – the myth that D2B consumers are superior, and that dumping on and viewing excluded people as inferior is ok, 4) for sexual reasons- they may be aroused by a person abusing an excluded, 5) It's a way for them to express and promote their viewpoint, and party-line to the excluded and those watching, 6) To confirm myths and lies that they have created or maintain: that 19 hijackers did 9/11, that Oswald killed JFK, that there is at least 1 God, that their excluded enemy is a pervert, gay, and/or has a mental problem. Ultimately shilling is because many D2B owners and wealthy people that fund shilling are just low-brow people. But while the people who fund shilling may be low-brow, the technology behind it is sophisticated.

Sophistication of the Shill: The D2B computers are smart and try to find words and phrases that a D2B consumer will feel comfortable taking money to say to an excluded. Many times the word will appeal partially to something they agree with, or that has a double meaning that satisfies two opposite sides. Mostly, though the shill always favors the D2B owners, and most of the time expresses a crude and/or distorted, dishonest, and deceptive view, which is the nature of buying a shill – the reason the D2B consumer needs to be paid is because they probably would not normally say the paid-for word or phrase. Shilling can be more than just a single word, for example, a person might be paid to say a complete sentence. Another tricky example is that a supervisor may be paid to harshly criticize an excluded employee under them, for example saying "what were you thinking?!" or "why didn't you do that?!", because it may lower the confidence of the excluded person. Sometimes people may be made to shill without being paid, for example if their muscles are moved remotely involuntarily, or when a shill word or sentence is remotely made to feel desirable to say. And the paid-for unwanted sounds to our ear don't stop when an excluded person walks away from a D2B consumer, because the shilling continues with paid-for audio sent directly to your ear neurons. Like a prank phone call, the audio usually takes a nasty and dishonest view, deifying the neuron, demonizing the excluded, deifying remote murderers, assaulters, molesters and liars, demonizing those trying to expose and punish them. Like a prank phone call, only the phone company knows who and where the sender is, and they don't tell the unconsensual receiver. An excluded can certainly try to think "no call list" or "out" in their mind, but it's probably useless; because for the neuron, money and their personal views are the bottom line, not majority views, fairness, decency, honesty, law, stopping violence, teaching science, etc.

Another aspect of the shill is the personality shaping that it does. Just like Pavlov's dogs, human behavior is shaped, not with food as much, but with D2BW money. D2B consumers are trained using fast payments of money to fulfill the desires of D2B owners and consumers. Not only are there money windows with offers to say rude words and phrases, but there must be a lot of "teleprompting" and "script reading" text. For example, perhaps phrases and words are offered by the neuron, just like the Internet, to help guide D2B consumers in their choices of what to say, what to see, how to win friends, how to deal with excluded people, how to lie about D2B, what to say to pretend that they know nothing about D2B or anything else, what to do, what to think, what opinion they should have, who to vote for, what products to buy, what songs to listen to, etc. – and then all with a little money "treat" if they obey probably. I've seen people young and old grab that $1000 or whatever in their eyes and just shout out words at me like "go!" and "perv!", in an airport, or walking around a shopping plaza- the Earth has become just like a virtual rude Nazi Vegas.

There is something so despicable and low-brow about tricking and abusing a deaf and blind person, it's like taking candy from a baby, and this is exactly what so many D2B owners and consumers engage in every day. They prey on the weak and those with the handicap (which they unfairly and needlessly create) of being excluded from D2B. D2B excluded people can't see and hear thoughts like they can, and just like deaf and blind people, excluded are easy targets who don't offer much resistance – they are the weak lambs of the herd. Remote neuron writing onto excluded people is very much like a rape because the excluded person does not ask for the writing, does not consent to the writing, and has never even been told that remote neuron writing exists and is being used on them.

Possibly one added aspect of the constant popup money window is to stop the D2B consumer mind from wandering or having time to direct their own thoughts.

One of the best ways of looking at those who routinely shill, is to see them as what they are: kind of like a robot that people put money into to echo statements out loud- and just to ignore them and what they say as quickly as possible; to see them basically as meaningless money machines. Some D2B consumers can give good advice, but the vast majority of money windows promote a slanted view. Obviously the best advice to D2B excluded people is not to look deeply into the words D2B consumers who shill a lot say. The more the excluded trusts the D2B consumer and views them as a "friend", the higher the money that is offered for the D2B consumer to shill and mislead the excluded person.

People are just like a share of a company stock; their value goes up and down. When they take money to shill, their value goes down because nobody wants to hear shilling around them. When I hear a person shill I many times think "hold out for the big money...that comes with honor, honesty, decency and integrity". Sometimes I think "you might as well put a negative sign in front of the money offered to shill, because that is what you are really getting".

The best thing about a clearly stated shill is that, you know, for sure, without any doubt whatsoever, that the person shilling definitely gets D2B windows. There is no question about it. They are not one of the millions of excluded victims that need to be told about and shown D2BW. That's a valuable piece of information for the D2B excluded. In fact shilling may be the best evidence for an excluded person that secret remote neuron reading and writing (and D2BW) even exists. Thank goodness for people's constant desire for money, because it provides some of the best physical evidence for RNRAW and D2BW. Excluded may even accumulate numerous videos of people shilling to present as evidence of remote neuron reading and writing in a court trial (with an excluded judge and jury perhaps), or to help educate and inform other excluded people about RNRAW and shilling. Another consolation is that they only lose money when they shill. They gain a short term thousand, but their credit goes down, and in the long term nobody wants to hire or date them because it's unpleasant to be with somebody who takes money to say abusive words. Sadly, those terrible D2B consumers who say unpleasant things for the idiot wealthy people who pay them, are being naturally selected for survival in being rewarded for abusing and misleading excluded people for money. On a positive note, any time a person shills at you, you can think that you might get richer, because that money may ultimately be transferred to the victim of the paid-for abuse. It's a tough life for many poor people and it's good to put everything into perspective, shilling is nonviolent and free speech; violence is the big evil, shilling is just annoying. Hopefully people will vote that all money gained from shilling will go to each particular victim.

In my experience, one characteristic of many D2B consumers who shill is that they can "dish it out, but they can't take it"- many times if ever I return a shill verbally, it usually has a dramatic effect on the person who shilled- they get very angry or upset. The key is understanding the dominant view of people who are being denied D2B windows by those who do receive D2B windows; those denied D2BW are viewed similarly to those accused of "witchcraft" and "blasphemy", or like native African people in South Africa under apartheid, Jewish people in Germany under the Nazis, or the "untouchables" of India. Each nation apparently has similar analogous groups; people who are viewed as second class for mistaken, petty, and/or sinister reasons. So a D2B excluded "talking back" as if they were an equal to a D2B consumer can be unusual and upsetting. The view is just like there is an "uppity" or disrespectful slave, but the truth is that the shill deserves to receive the same abuse that they deliver to their excluded victims- most of whom are excluded not because they are somehow flawed or inferior, but probably much of the time, because of their unwillingness to lie about D2B, about scientific truths, or for their criticisms of popular mistaken religious beliefs. The D2B included can be shockingly rude; they harass and bully, many times receiving thousands of dollars in the process, but if a D2B excluded ever fires back, or "returns a shill", all the blame is put on the D2B excluded. It's a lot like a racist saying "did that slave talk back to you?!".

As a D2B excluded victim of shilling, I encourage you to be better than the people that shill, to rise above, to not become angry or violent. Just like any victim of a corrupted "class" societal structure where fine and wonderful people are treated like dirt, don't sink to their level, instead, focus on your goals and dreams. It very well may be that every single penny (ESP!) gained by those who shill may someday be paid to the victim of the shill. And when I say that you, as a D2B excluded, have to be better than the D2B owners and consumers, I mean that you have to be 1000 times better, because just like in any place where there is an unfair segregation, the second class must always be 1000 times more friendly, talented, and productive to get a job, a room, a mate, etc., even though they get 1000 times more abuse, and 1000 times less information, physical pleasure, reproductive rights, money, job, and housing opportunities. Plenty of wonderful, beautiful, brilliant, skilled, and talented people have been burned at the stake, gassed in death camps, galvanized, etc., because they were labeled second class citizens by the dominant uneducated wealthy D2B big-lie murderer elites.

Sophistication of the remote neuron reading and writing technology

If remote neuron reading and writing is 700 years old, then clearly those who own and control it have had a long time to develop it, and it shows. There are many aspects that, as an excluded, I certainly am unaware of. But, one thing seems clear, and that is that the owners of RNRAW have computers that can draw (render) many objects in three-dimensions and in real-time. It seems likely that all 7 billion people are modeled in 3D and their motions predicted. The computers probably predict the motions of humans and other 3D objects generally over a long term- with regular activities like when we will go to our job, etc. – but also into the near future, perhaps even hours into the future – for example knowing when we will reach any particular stop light, etc.

Defend yourself against the remote neuron writing

For the love of life, defend yourself against biting on bad suggestions. These remote neuron writings can take the form of barely visible images sent to your thought-screen, barely audible sounds sent to your thought-audio, and memories that are triggered in your mind – there are many subtle and sneaky remote neuron techniques. I know from experience that sounds remotely sent to the ear and thought-audio of a D2B excluded can be misinterpreted as "the voice of God". I made this mistake – and I'm not religious! Think of how common the phrase "heard voices in their head" is. Many times, this is what those who have murdered somebody say. Mark David Chapman said how he heard "do it! do it!" in his thought-audio before he killed John Lennon. Recently a person jailed for the murder of homeless people, slammed his head against part of his cell trying to make the voices in his head stop. You can see how a person who sends suggestions to a D2B excluded person might claim "I never pulled the trigger" and plead "free speech". Getting an excluded person who owns a gun or other weapon to do violence is a classic remote neuron writing. Much of this shocking and disgusting violence and abuse of excluded pawns would end if only the public was even just told and shown the details of remote neuron writing.

One possibly helpful idea is to "fire back" in your mind at the remote molesters. I find myself at the "fire back" laser in my mind every 5 seconds. Imagine those that don't fire back, but readily accept the terrible remote neuron written sounds, images and other suggestions- they are easier targets and just like typical predators, the neuron rape writers pray on the weakest of the herd. The feeling of "stage fright" can be remotely sent. Even simply "fear" remotely stimulated in your mind can be remotely sent – for example when you turn out the lights before going to bed. I find that firing back in my mind helps tremendously. Let's hope and vote that the remote particle molesters must lose their neuron writing permission. In addition, they should lose their neuron reading permission – at least until neuron reading and writing is public and their victims can read and write from and to their neurons too. Certainly I think that their particle molestation should be done back to them. One idea is to think a response that is similar to the molestation being done to you- if you are being made to feel like you are getting an enema, think about them receiving the same sensation. I think the key is not to become angry and in particular not to become violent. It may be that the images and sounds victims of remote particle molestation think in response have almost no significance at all and that much of remote molestation may be to try to make a person "blow a fuse", become violent, and attack people who, unlike the remote molesters, *can* be seen and *are* within physical reach of the victim.

You might think that a wealthy people that own and control remote neuron reading and writing would be "good parents" to the excluded "children" of Earth, that those with the immense power and advantage of remote neuron reading and writing, would at least, since they refuse to openly describe and share the technology, show responsibility, and be like a helpful "seeing-eye dog" to those excluded they have made "blind". But experience has shown that the exact opposite is closer to the truth. If they are a parent, they are a parent who constantly tries to find new ways to kill their kids. If they are a seeing-eye dog, they are a seeing-eye dog with a death-wish for its blind owner. The focus of most remote neuron writing is on exterminating, assaulting, molesting, and misleading most of humanity.

Figure 2. What looking at people with Direct-to-Brain Windows enabled might look like. Notice that the dog (that receives D2B windows) see's part of it's nose on it's eye-screen. Notice that the only people with friends or babies are D2B consumers.

Included and Excluded

There is clearly a segregation of those who get direct-to-brain windows (many times called the "included") and those who have never even heard of direct-to-brain windows or remote neuron reading and writing (the "excluded"). There will always be inequality among humans; those who have more money, those who are younger, better looking, smarter, etc. So in some sense, for a D2B consumer walking around with D2B windows seeing many other people without D2B, it may be like the way those with homes look at homeless people – with pity, but an acceptance that some people are always going to have less and have a tougher life than we do just like others have more, and have an easier life than we do. But the D2B segregation is more like the segregation directed at Jewish people in Nazi Germany, or African American people in the United States under slavery; it goes far beyond just a difference in income or appearance, and is, I think, in large part, an "ideological genocide", based on a person's religious, and political beliefs, how honest a person is, and probably to a large extent their gender and race. Then to not openly acknowledge this massive secret system, even to the extent of not allowing a D2B consumer to admit that they receive D2B (even to their own children), is clearly evil and unnatural. One of the worst aspects of the D2B segregation is that, unlike the openly debated segregation based on gender or race which is often documented publicly in laws, like the famous Nazi laws forbidding interracial marriage, it is almost impossible for the D2B excluded to "see on paper", or to even know that they are being extremely discriminated against, and being remotely abused.

Denying millions of people knowledge and regular use of D2B creates an imposed disability and handicap in millions of otherwise healthy people

The owners and controllers of the vast nanometer sized particle (neuron) technology have created a handicap in denying many people the regular service of direct-to-brain windows. Because so many millions of people do receive regular D2BW, it creates a disability for those who are denied- because modern people are communicating through thought-audio and sending thought-images back and forth, but the excluded must be told out loud and have a physical image shown to them to understand what has happened. D2B excluded walk around and grope in darkness while the D2B consumers just watch and worry what the poor excluded people might do. That excluded people are not even told that many people see and hear their thought images and sounds, and that even telling them this out loud is forbidden by the neuron owners adds an extra shockingly callous and inhuman dimension to the imposed disability. I can understand denying people with multiple counts of violence the right to receive D2B, and removing the D2B write permission for those with multiple remote particle molestations, but most of the people being excluded from using the "thought-phone" or "thought-net" haven't committed any violent crimes. The popular view among D2B owners has traditionally been one that favors an ignorant, uninformed, deliberately deceived, and uneducated public. But, I think the opposite view, informing and educating people, is the path to safety, survival, and to physical and intellectual pleasure. Knowing that our fate is to try to build a globular cluster, serves as an incentive, helps people to see the big picture, and they then shape their lives with that in mind. Probably many mistaken scientific beliefs and barbaric religious traditions would come to an end if everybody learned the truth about RNRAW. Certainly many people denied D2B would then be better prepared to reject bad remotely neuron written suggestions. Everybody knowing that thought can be seen and heard would vastly reduce the needless use of paper and ancient technology, in addition to the time saved in not having to repeat everything out loud again. The free flow of information always helps because the majority can see what the problems are and work to solve them, as opposed to groping in darkness with mistaken beliefs while those causes of damage continue to grow unstopped. Beyond that, communicating through thought-audio and thought-images has become so common that it should be made public, like the cellphone, and like a cellphone and the Internet, middle income people should be able to use D2BW for a low monthly cost, in addition to receiving a percentage of the money paid by the public to the telecom companies to see them and their thoughts. We're hurting ourselves by keeping remote neuron reading and writing a secret because analogous species of other stars are not delaying, as they continue to grab more stars.

D2B consumers have to pretend

One D2B consumer kindly said to me "We have to talk", which is a very important truth of this dark and medieval era- D2B consumers can just think to each other (sending thought-audio and images back and forth), but have to talk out loud to D2B excluded people. If everybody could regularly see and hear thought, obviously there would be no need to repeat thought-audio or describe images people see in their eyes or thoughts out loud. Making D2B consumers have to talk out loud is a big waste of time and very annoying. The current rulers of the neuron insist, as generations before have, on maintaining the myth that D2B consumers cannot see thought-screens and hear thought-sounds. For example, if there is a meeting at work (fig. 1), and all the participants are D2B consumers, they still MUST talk out loud – they cannot simply conduct the meeting in thought audio and image. Is that not absurd? You can imagine society on Earth if and when the segregation is public and D2B consumers hold meetings in silence, occasionally talking out loud for the benefit, or to explain something to a "deaf" excluded person. Perhaps the excluded people will wear headphones and the D2B consumers will type their thought-audio notes into a synthetic voice machine for excluded people forbidden from receiving D2BW to hear with headphones.

D2B consumers have to constantly "play dumb". They know everything about the excluded, but when the excluded asks "did you know this?" the D2B consumer must always pretend to be surprised and have no knowledge of events at all. The "playing dumb" goes to a ridiculous extent, comically parodied in the movie "JFK II"[2]- I can imagine D2B

---

[2] John Hankey, "JFK II" , 2003.
http://topdocumentaryfilms.com/jfk-ii-the-bush-connection/

consumers saying "what's a neuron?", "what's a particle?", "what's a JFK?"... just endless ridiculous denial of obvious truths they all know in great detail.

One great annoyance is that D2B consumers cannot respond to the thought-audio of a D2B excluded who directs thought-audio at them. The D2B consumer cannot answer any thought-audio questions out loud, but some may nod or make other subtle gestures or use coded words to respond out loud.

Who gets D2B and who doesn't? Who is included and who is excluded? Who is "innie" and who is "outie"?

One of the classic questions of these centuries is: "Is so-and-so included?". For example "Is my parent included but I'm not?", and "Is my child included but I'm not?". It's unbelievable, but yes, in many families, one or more members may be excluded from even knowing about D2B, while others receive regular D2B windows. In my own immediate family both my Dad and brother are "included", and receive some regular D2B windows (although as excluded, you can never be 100% certain), but my mother was never included. Those family members who do get D2B are absolutely required to lie if ever asked if they see or hear thoughts. However, many included family members do care about excluded family members and so are constantly "hinting". Most of this hinting takes the form of key words said out loud or in text messages. Hinting is acceptable, but any kind of direct and open acknowledgement of
getting D2B is absolutely forbidden and must be severely punished – probably by losing D2B service for long periods of time, and perhaps permanently. An included uncle of mine hinted to me about remote neuron reading when I was a teenager- only much later in my 30s did I realize the importance of what he was telling me about pilots controlling a plane using only their thoughts. For excluded people that know they are excluded (that "try to use the phone" as one D2B consumer put it nicely- they direct thought-audio at people), hanging out with D2B consumers, in particular D2B included family members, can be a tough situation, because the excluded who knows about D2B usually trusts their family members (unlike many of the people who shill

around them) to tell them the truth, and tries to get some answers from them. D2B consumers see so much more than excluded people and so have a much better view of the big picture surrounding an excluded person. As a D2B excluded that knows about D2B, all I do is think thought-audio questions to my included relatives, and all they can do is just hint by nodding and hinting out loud by saying "**yeah**...I think this or that", or "You **know**, I ...", and by using other creative word choices. Once, for example, I asked in thought-audio – "What did Mom think about a lot?" and the answer: "mao"- which implies lots of images of food (eating food is sometimes called "mowing food")- my Mom loved all things food, and plus there is a double-word score with the word "Mao" for "Mao Zedong", the famous leader of Communism in China; being atheist, as my Mom was, instantly earns you the label of "Communist" even though my Mom was outspokenly for Ayn Rand style "laissez-faire" free market capitalism- she had never even heard about *full democracy* where the public gets to vote on the laws they are subjected to. Be aware that, even your immediate family members will deny and lie about receiving D2BW to you if confronted directly, in particular about seeing and hearing your thoughts. Whatever makes them lie is apparently a very scary threat or something that would make any average person lie to even their own mother. In addition, there is the addiction and absolute fear of losing D2BW service. There is a clear analogy to being locked in a prison camp: pretty quickly, you realize that to survive you need to say whatever the camp guards tell you to say no matter how untrue.

Are there any guides or techniques a D2B excluded person can use to determine who is in and who is out? Believe it or not, there are. If a person "shills" you know they are "included" (they receive D2B), but you have to really hear a number of clearly spoken shills to be sure. This is one really good reason that an excluded person should not become angry or emotional when hearing a shill, but instead, recognize the great value of knowing that a person gets D2B. If the shills appear to be very quick involuntary "Freudian slips" – then the person is probably excluded too – because a shill takes preparation, where an involuntary motor neuron write is too fast to be pre-planned. Some D2B verbal statements for money are hard to detect, because the D2B consumers definitely want that money, but they definitely don't want to say something offensive or mean to the excluded target. The shill payers definitely want to direct the topic of the thoughts of excluded if they can, and so the shill-for-money words may often seem somewhat neutral. Terrible D2B consumers are easy to detect, because they always take the money to say clear and terrible insults, but decent D2B consumers will usually let you know that they see with some kind of words, in particular like "I see", or a gesture like holding a fist or finger to their mouth or eye. The absence of a shill is one of the best indicators of an excluded. Think of those around you who have never shilled- they are most likely excluded. The only other alternative is that they are D2B consumers that refuse to take money to say things to excluded people. Most D2BW consumers say something or make some clear gesture which reveals their D2BW reception pretty quickly. If any person deliberately covers their mouth with their hand, or makes some kind of deliberate and meaningful hand gesture just when you first see them, you know instantly that they are probably D2B consumers. The number one give away of a D2B consumer is the shill – it's driven by money and so many D2B consumers bite on them. In fact, the excluded may form a searchable Internet database or "Wiki" that identifies each human of Earth by name and photo, and provides video evidence of shills and gestures, or from thermal imaging cameras, that indicate if each person is a "D2B consumer", or a "presumed D2B excluded".

Some people may nod their head in response to your thought audio question. If they do then probably they are a D2B consumer, if they don't, then they are either too scared, don't care enough about you to respond, or are excluded. D2B computers are probably set to routinely cause remote control nods in excluded people in false response to thought-audio questions from excluded people that know about the massive D2BW segregation, using the D2B owner's already extreme advantage to fight against excluded people even knowing who is excluded, informing each other about RNRAW, connecting, or uniting.

A D2B consumer will generally not look at an excluded, while an excluded probably will look at another excluded if passing them. A D2B consumer doesn't have to look at anybody around them, because they get D2B windows with real-time video from nearby floating wireless nano-cameras, but D2B excluded have to look around to see the people around them. In addition, a D2B consumer usually wants nothing to do with an excluded person, because communication is impossible, because there is no chance to get anything (like money or a job) from an excluded, and because there is a fear of an excluded acting out remotely written bad suggestions. So a D2B consumer will not make eye contact with an excluded because 1) they already see them in their eyes, 2) a relationship would be impossible and annoying, 3) there is nothing an excluded person can offer them, and 4) they fear excluded people as dangerous and more likely to act on bad remotely neuron written suggestions.

Is one of my friends included? Mostly D2B consumers only befriend other D2B consumers, but they may have a "hi/bye" relationship with an excluded. It's very tough for excluded people to make friends. If they have a lot of friends, probably they are a D2B consumer. If they are club leaders, event organizers, etc. – probably they are included because they can easily and rapidly contact and communicate with many team members. In addition, D2B consumers are in contact with all the "higher ups", and so can quickly get approval or denial from those above them.

One method is just to ask people directly: "Do you get direct-to-brain windows?" and then give them a few seconds to answer. Believe it or not, in my experience, some D2B consumers will actually admit that, "yes", they do get direct-to-brain windows. I can say that at least four people I have asked verbally admitted it. It helps to ask, but be careful not to get into trouble with your job, family, or friends, etc. Once, I was confronted by a supervisor because I told a coworker that some people might be able to hear thoughts. Knowing if a person gets D2B, I think, is the first and probably most important thing an excluded should want to know about anybody, because it makes a tremendous difference. As an excluded, if you know that somebody gets D2B, then you know that you don't need to tell them anything, because they already know it all. Asking if a person gets D2B is mostly a win-win question, because if they do get D2B, then they already know everything about you, there is no harm done, and you are helping to make talk about D2BW more common and acceptable, but if they don't get D2B, then you are telling them about something very valuable, that they may think about for the rest of their lives, and that may help them tremendously in defending against those remotely writing on their neurons. A more subtle way of asking is "Have you ever heard of direct-to-brain windows?"- then you don't sound as crazy if they are actually excluded. It's good to hear what they say, because that represents part of the current public view of D2BW.

Those who get D2BW prefer not to talk about D2BW and RNRAW, while those who are excluded show little or no fear in talking about it.

There are other common characteristics of included and excluded. If there is no clear shilling, and no admission of "I see", then you are left with a person that is either an excluded, or a D2B consumer that is too afraid to admit it. Generally, excluded probably are mostly: males, young kids, elderly, childless, liberals, democrats, non-white, obese, smoke, use recreational drugs, drink alcohol excessively, atheist/agnostic/not-religious, or very religious, poor, not beautiful, have small breasts, have some kind of learning or neurological disability, have done unusual things and got locked in a psych hospital (are "crazy"), are homo- or bi-sexual or transgender, are overly honest, are against restrictions on info and secrecy, care too much about other people, or are overly or openly sexual- are "sex offenders" (note that "are violent" doesn't appear to be an issue but could be). It's tough to be sure, one D2B said "there's no rhyme or reason to it", and probably that is true to some extent. Probably a large part of who gets D2B and who doesn't is simply determined by if a person's parents are included. There may be lineages of families where some have been D2B consumers and the others D2B excluded for hundreds of years.

One tricky aspect is that excluded are routinely made to "mimic" included people. Because anybody can have their neurons remotely written on, an excluded may "sound" like an included by 1) echoing your thought audio, 2) by saying something that makes them seem like they saw something you did inside your house, 3) by apparently "shilling", and 4) by making excluded people gesture or nod through remote writing. If a person says a shill that is extremely fast – too fast to be thought out – it was probably involuntary remote neuron writing. It's very tough to quickly replace some word like "think" with "fink" for example, although many insiders know these phrases, get the words tele-prompted in front of their eyes, and can quickly get the shill money window. If a person's apparent response to a thought-audio question involves them itching, then probably they are excluded.

Another possible test to see if a person is excluded is to ask them if they know that 9/11 was a controlled demolition. If shocked, embarrassed, and very doubtful, probably they are excluded, but if they already know or actively provide arguments against it, probably they are D2B consumers.

If they have reproduced, probably they are included. I went to a high-school reunion and most of my friends had kids that were about 5-7 years old, and so it's likely that most of them got included around 5-7 years before that reunion, which would put their D2B inclusion date around age 30.

Many D2B consumers say hints that reveal helpful info about who is excluded and who is included, for example some have used the phrase "flat out" which implies that females with small breasts (a "flat chest") may more often be excluded, while women with large breasts may more often get included, in particular at the request of many of the wealthy D2B "johns".

Probably many non-white people are excluded. Every time there is a murder that makes it to the news, probably the murderers and victims were excluded non-white people. The murderers are most likely poor non-white excluded people who followed some remote neuron written suggestion to do violence.

A very tricky aspect of D2B is that most D2B consumers, apparently, get a very limited service. For example, you can't presume that just because a person was the victim of a crime, even a violent crime like murder, that they were a D2B excluded. For example, the Kennedy's and many of the victims of 9/11 were probably D2B consumers who didn't get the D2B windows that they needed to see in order to protect themselves from their attackers who many neuron owners and probably even many D2B consumers certainly knew about.

Probably most excluded people are poor people. However, I know of many poor people who do get D2BW, in particular grocery cashiers, because many of them routinely take the money to shill. But the D2B service for poor people must be very limited, because everything costs money - that's why so many poor people shill. Most wealthy people probably don't shill because they don't need the money and it makes them look cheap and low-brow.

Probably many atheists are excluded. In the USA, atheists, ironically for so modern a nation, are less popular that almost all other groups. It seems likely that remote neuron writing is mostly owned and controlled by those who believe in one or more Gods, and in particular (in America, Australia, and Europe) "followers" of Jesus. Perhaps the majority of those who are D2B consumers were "gotten" (included into D2BW) through people in their church, mosque, synagogue, or temple. D2B owners are probably most comfortable with people who follow the herd. My mom was openly atheist and she was excluded all of her life. Speaking from my own experience as the child of an atheist who later became an atheist: if you are an atheist male, put your thumb and first finger together for your daily masturbation, and get used to it, because that's what the Church has planned for you, and they use every dishonest trick in the book, including remote neuron writing, and violence, to do it. That's also about as close to seeing a direct-to-brain window as you will probably ever come. Simply put, those people tricked by the lies and supernatural claims of the religions who own and control most of D2BW and RNRAW don't want the enemy, those who tell the truth, and who are supporters of science and atheism, to make more little heretics.

In the USA, under the current Bush legacy (where the Kennedy's, MLK, John Lennon, and other popular democratic leaders are all murdered) probably Democrats, progressives, and liberals tend to be excluded. It seems likely that the majority of D2B is probably Republican owned. They are in the dominant position and don't want to help people on the other side who might help to bring them down, or get them jailed for their involvement in violent crimes.

Teachers, engineers, and doctors tend to be excluded. My dad is a doctor and he was excluded for decades. Teachers are the focus of many sex-with-minors investigations which implies that they are excluded, because if they were D2B consumers, they would know better than to try to sexually touch a person under the age of 18- and have plenty of opportunities for physical pleasure with the many other D2B consumers at the drop of a thought. I'm an engineer and I am excluded. Probably many architects and engineers in the group: "Architects and Engineers for 9/11 truth", are excluded, because not many D2B consumers (who all know that 9/11 was controlled demolition) would feel brave enough to publicly say so and risk losing their D2B service because of violating the code of secrecy. It seems pretty clear that, as was the case for the hording by Pol Pot and a few others in Cambodia, educated people are the targets of the crude and uneducated D2B owners and consumers who remotely write on them, and who haven't done a lot of work in their life.

It may be that excluded do a lot of work, like worker-bees or worker-ants, so perhaps those with tremendous working achievements were excluded. Included D2B owners and consumers, find themselves in a role similar to a queen bee. The argument may be: "we don't want to interrupt their important work by including them", "we don't want to slow or stop their contribution to progress for the excluded and humanity" (as opposed to doing something good for humanity as D2B owners and consumers themselves like going public with remote neuron reading and writing technology). Many D2B consumers must think "maybe the excluded is smart enough to figure it out, and tell the others – if we keep them excluded they can help other excluded people by saying what we can't say", and "maybe the smart engineering excluded can duplicate some of this secret technology and therefore make it public so everybody can enjoy it publicly". Certainly, when a person gets included, they probably instantly get dates, get to make a family, and as a result, their output of non-family related work probably is greatly reduced. But the truth is that the entire system of purposely excluding half or more of the planet from a regular direct-to-brain windows interface is simply a vicious and unnecessary system to continue.

I often walk around and look at all the people around me and wonder who has D2B windows and who doesn't. It must be a sad feeling for an included person to look at a person or group of people and see that they are "excluded"; without any D2B windows, and with only their two "eyes" and "thought" windows above them. Even as an excluded it shocks me to realize that some of the people around me are looking at D2B windows and that others have never even heard of direct-to-brain windows. It must be sad to see people who are without D2B windows, because you know instantly that they are going to have tough lives – constantly misled by remote neuron writing, denied jobs, any physical pleasure and sex, friends, etc. But, like so many classic examples of history, probably many D2B consumers get familiar with seeing the many excluded people, and become desensitized to this senseless segregation.

So that is an important question of our time: "Who is excluded and who is included?" As a subset of that question there is the question: "Who is excluded that knows that they are excluded?". Of the famous people of the past: were they excluded or did they actually receive direct-to-brain windows and pretend that they didn't? So far, I know of no famous person of history who, if they knew that they were excluded, ever left any explicit public record of it. The closest is perhaps William Crookes who said telepathy through technology may be possible. Was Crookes a super-smart excluded or an extremely compassionate-included? Another is Hugo Gernsbach, who also wrote about recording thought-audio in 1911. Was Gernsbach a brilliant excluded or overly honest included? Who knows how many of the great thinkers of the past were excluded from D2B windows? Judging just from people who did not reproduce, perhaps even famous humans like Isaac Newton, and Beethoven were excluded all of their lives. It seems likely that victims of murder and other violence were certainly mostly, if not completely, excluded, or else they would have received a warning in their D2B windows. Even popular celebrities like John Lennon, RFK, MLK, JFK, Nicole Simpson, Gabrielle Giffords, the murdered kids like Larry King, Chelsea King, Adam Walsh, Jessica Lundsford, Jon Benet Ramsey, those kids recently murdered in Norway and their parents, all obviously didn't see the D2B videos of their attacker and their thought screen- which many others must have seen and been watching before, during, and after the violence. One shocking truth that no excluded can deny is that serial killers are allowed to continue their murdering over many years, even when all the D2B owners, and many millions of D2B consumers must definitely and positively know who they are from even simple street cameras (most killers use a car), if not from the killer's memories which regularly appear on their thought-screens. Not only did their victims and the loved ones of their victims probably not see the live image and thoughts of the murderers, but they probably don't get to see the live image or thought-screens of the people who wrote to the neurons of the murderers which possibly influenced and/or suggested the violent crimes that were eventually committed.

From my own experience and the many clear examples of history it is pretty clear that the rule has been for centuries that some of the worst and grossest people get D2B while many of the most decent and wonderful people are excluded, and this is a reflection of the reign of religious barbarism. Murderers, torturers, remote molesters with thousands of counts, liars, all get to see, while their gentle and honest victims don't get to see, and are constantly remotely abused; just absolutely upside down.

The top 4 daily reminders that a massive and highly developed secret remote neuron technology exists

4) Being remotely made to itch
3) Having a muscle remotely moved
2) A D2B consumer shills
1) Hearing remotely written sounds

How many humans get D2BW and how many are excluded?

This is another classic question. As excluded, of course, we have to guess, and hope D2B consumers will hint to us. One thing that is clear is that there are many people who get D2B and many that do not. Perhaps the split is 50/50. But probably, given the difference between developed and undeveloped nations, the ratio is probably more like 3:10, 3 of 10 humans receive D2BW. Many are included, because, in my experience, in Southern California, the vast majority shill in some form or another- those that do not shill are extremely rare- maybe 1 in 30. But yet, clearly many people are excluded because they fill movie theaters, buy newspapers and magazines, books, and watch television. It's tough to know if these "book buyers", etc. are D2B consumers that are somehow doing obsolete things. In addition, in order to communicate D2B consumers are absolutely required by the D2B owners to use the old technologies (like talking out loud, telephones, email, etc.). For example, I see thousands of people, all the time, talking on cellphones, but most of them are probably D2B consumers talking to other D2B consumers, even though a cellphone is completely unnecessary for D2B communication. It's tough to know for sure. One D2B consumer said to me "there's a lot out there", which I think is probably true.

What about the young children of D2B consumers, do they get D2B too?

I don't know and maybe everybody is different – even some pets may get D2B – and be able to communicate through thought-sounds and images. One person hinted that some thought-audio talking does get sent from parent to child – at least when they are very young- perhaps thought-audio like "are you hungry?", but another D2B parent said flatly "we want to tell our kids". My Mom was definitely excluded from birth to death, and probably her parents (and all earlier ancestors) were also excluded.

"Excluded orphans"

Most D2B parents probably constantly communicate with their D2B kids (strictly through thought) while their kids are growing up, in high school, in college, and after. But, D2B excluded parents with kids, have to use the Internet, cell phones, etc. to see and communicate with their kids. So, unlike D2B kids, the D2B "excluded orphans", those who are excluded, and whose parents are excluded, are the target of D2B abuse. If something terrible is happening to a child, the D2B excluded cannot quickly put an end to it, and may never know about it. D2B parents get to see life from the eyes of their child and see their child's thought screen and so know what the kid is thinking of, and can quickly guide the child into making good decisions and away from bad ones, but excluded kids, have only D2B writers to "parent" and guide them.

In particular, without any doubt, and from first hand experience, many of these young excluded kids are the target of sexually inappropriate D2B suggestions. Because they are "D2B orphans" (are excluded and have excluded parents), they are easy targets of the many criminals who buy and sell remote neuron writing transactions. Probably many images sent to excluded, are an effort to try and fulfill the sexual fantasies (as seen on their thought-screen during masturbation or sex) of high-paying D2B customers. This, many times, involves sending excluded people direct-to-thought suggestive images of same-gender touching, of them in opposite gender clothing, of them nude, masturbating, or having intercourse in public, or with other species, etc.

Parents should actively inform their kids about sex, I think, to reach them before the remote neuron writers do. People need to accept publicly and openly that the vast majority of young people (after age 10) masturbate every day. It's unhealthy, neglect, dishonest, and idiocy to deny this fact and keep it a secret. We are all products of sex. The current system of denying and neglecting young people's right to physical pleasure is inhuman and shocking. Pleasure and sex education should be a large part of a young person's life. There is a natural desire in most young people to feel warmth and affection. They should be allowed to kiss and fondle each other with clear and constant boundaries and consent, and with the parents being fully aware and helping to actively guide their child's choices.

In particular excluded families should be vocal and active in advising their kids in terms of dating, kissing, masturbating, remote neuron writing, and the laws concerning sex, because even if you do inform and guide your child into regular and healthy kissing and fondling, and daily masturbated orgasm, the D2B writers are going to shape a large part of what guides an excluded child's daily desired physical pleasure decisions. Acting out some of the sexually inappropriate suggestions (like masturbating in public, doing something physically pleasureful with a child they are baby-sitting, or a pet dog, etc.), even though non-violent and many times natural, might haunt the excluded child for the rest of their life, in particular in the bizarre violence-tolerating, pleasure-forbidding society we live in.

The D2B included have an extremely better opportunity to protect their kids, and can interact with their kids through D2BW every second of the day, while being excluded from D2B is to be an orphan with absent parents, and a favorite prey of wealthy D2B criminal owners and consumers. The 9/11 controlled demolition, JFK, RFK murders and their associated cover-ups that still persist to this day are clear examples that laws, even laws against violence, don't apply to most D2B owners and consumers.

“Pedding”, “perving”, “sex criming”, “insanitizing”, “thefting”, and “violenting” D2B excluded through remote neuron writing

Wealthy D2B consumers and owners have learned from years of experience, that one very effective way to lower the value of their enemies is to use remote neuron writing to get D2B excluded people to commit crimes and do unpopular things. For the many wealthy religious D2B consumers and owners, the targets are probably often liberals, scientists, and non-religious people. For racist D2B consumers and owners the targets are mostly non-white people and those who are supporters of racial integration. It seems likely too that liberals, scientists, and non-religious people may use remote neuron writing to try and lower the value of popular violent or religious excluded using similar remotely written suggestive images, sounds, muscle moves, and other neuron activations. Much of this writing may even be computer conceived and controlled- being somewhat faster and planning farther into the future than the average human brain can. Every second millions of dollars in remote neuron writing transactions occur with just this effort in mind.

One major goal is to make a person do something sexually inappropriate with another person under the age of 18, for example, by sending the excluded many quick images (“flashing them”) with well rendered suggestive images of a hand touching a butt, or a kiss, etc. they can get the excluded to decide to act out that remotely written suggestion- to think “will I kiss that person”, or “what would that be like?” – without realizing that – without realizing that they are always on “candid camera” (at least for the many millions of D2B consumers) and that these evil remote neuron writers are trying to ruin their career, popularity, and reputation. Even to get a person to think of images of young people during masturbation may be enough to mark a person as a "pedophile" or "sex offender" for life, because many millions of D2B consumers can refer to the thought images of the excluded person as evidence of pedophilia.

There are numerous techniques of remote neuron writing, and this book is only scratching the surface. Another method may be to remotely molest a D2B excluded person to make them angry or violent, for example by making them constantly itch, or sending unpleasant and upsetting sounds to their thought-audio (like their mate having sex with their enemies or friends). An angry person is unattractive, and appears to be a dangerous person. Being angry makes a person look bad and doing violence really lowers the value of an excluded person, because D2B consumers will not want to hire or befriend a person who is easily made violent.

Getting an excluded person to do something that will get them locked in a psychiatric hospital is a very popular form of neuron writing. Just getting a person locked in a psychiatric hospital once is enough to forever label them as insane, unstable, unpredictable, and dangerous. It's probably easy to get people to make a public disturbance that gets them locked in a psychiatric hospital. One method is to use remote neuron writing to make an excluded person think that God is ordering them to do unusual things, or that they are wildly famous, like a celebrity, or good looking, or extremely wise, and are soon going to be very wealthy. The excluded doesn't realize that they are better off focusing on making sure they have a place to live, on getting a date, a kiss, making a family, etc.- because none of those remote neuron writings ever come true- in particular for excluded people. Some part of the trick is that the excluded think that we live in a high-minded and decent society, but the reality, that they don't realize, is that we are living in a dark age where secret violence, lying, and remote trickery are the rule, and where truth, honesty, and decency are the extreme exception, because of the rise of Christianity, and the 700 year-old secret of RNRAW and D2BW. Even the opposite technique is sometimes used, to make a person think that they are of very low value, that they are viewed as perverted, insane, ugly, obese, stupid, old, etc. and then the excluded may be more easily persuaded to do something desperate or to settle for a very bad deal. Some D2B excluded, who realize that many people watch them may be easily convinced to do unusual things to "please the audience", or "have a laugh" with their invisible one-sided audience.

The D2B consumers and owners know that getting their enemy to be thought of as a "sex offender", "violent", "unstable", or a "thief" is a good way to keep them excluded, to lower their popularity and influence, and to restrict their potential wealth, job potential, and in particular their chances of reproducing and making more enemies.

Lots of D2B consumers constantly view me and other excluded people as "crazy" and "unstable", but in truth, and in my experience, as an excluded you never know what to expect from D2B consumers, they flip-flop all over the place, many times for D2B money- one minute they are friendly then the next minute they are mean- but if there is a consistent mood it's mean and sour- certainly for those who fund the constant nasty shill.

There is no question that many methods of remote neuron writing being used are complex and advanced. Much of the neuron writing is somewhat difficult to explain with words, in particular from my very limited excluded position. Beyond that, ultimately, it's clear that if a remote suggestion isn't enough, remote neuron writing can definitely be used to simply move the human mind and muscles any way that is desired. But isn't it interesting how the current system has some kind of planetary unwritten code where writing quick millisecond suggestions (no matter how violent or terrible) are routine, but long periods of continuous remote neuron writing are apparently very rare. Remotely holding lung muscles for a long duration to suffocate a person may be one exception where remote writing is not interrupted for a few minutes.

D2B thought feedback is very helpful

Instantly seeing the thought screen and hearing the thought-audio of a person allows a D2B consumer to quickly see where they may be going wrong, and to adapt. You know exactly what the person you are trying to befriend or date wants, likes, hates, etc. For example, you know they love some particular musical group so you can buy them tickets to see that group, or they like some restaurant, so you can invite them there. You know that they expect you to wear formal or casual clothes, etc. while the excluded knows none of these things. Just like the panty-liner ad on the timeline, the D2B consumer can see a picture of the potential mate kissing them on their thought-screen and know that they are thinking about a kiss – while an excluded would have no idea at all. Also as a D2B consumer, you may know if a person around you is violent, has a weapon, has a violent past, or doesn't like you and you can avoid them. But excluded people never know if a person near them is stalking them, has a weapon, is planning to do violence to them, or hates them; an excluded person is constantly walking through a "mine field". In particular a D2B consumer can use their D2B advantage to know where an excluded person will be, and so can more easily stalk, threaten and attack them. In late 2004, some D2B consumers put dead rats on the bike path I regularly used and near my driveway (fig. 3). Some D2B consumers recently exploded firecrackers around me, and I'm sure, often do many other such activities that require knowledge of where excluded people are located that they can only get from the D2B network. It's like something right out of Nazi Germany, but far more subtle, hidden, and protected by secrecy. The Safeway Giffords shooting, the Kennedy killings, 9/11, 7/7, the people recently murdered in Norway, and similar kinds of murders show clearly that many D2B consumers are unaware of violent people around them, presuming that some of those victims were and/or are D2B consumers, which seems very likely.

Figure 3 Left: A dead rat in an "H" shape on 10/16/2004 with a close-up. Right: A dead rat on the sidewalk next to my apartment on 12/12/2004, just before and after the US election. Coincidence, or stalking and threats of violence by people using a D2BW advantage?

The phony D2B tricks

So much of life on Earth under the D2B secret is phony. Not only do D2B consumers have to talk out loud and pretend that they don't see and hear thoughts during meetings, and even to other family members, but there are millions of other examples of the constant fakery. Here are some:

- Televised "spelling bees" are all fake: only people that receive D2BW can be contestants, and what they say is all scripted and sent to their eyes using D2BW.
- Television "game shows" are all fake: all contestants are required to be D2B consumers, probably because the show owners prefer to have a lot of control over the content and to try and maximize the number of viewers based on audience preferences.
- We have to accept that any sporting event must be subject to remote neuron writing and the free and fast money market transactions of D2BW. People can easily be made to drop the ball, etc. with remote neuron writing. RNRAW has ruined sports in my view because now sporting matches are more a contest between remote particle device owners than between skilled athletes. Even without RNRAW, I find watching sports dull and meaningless, in particular knowing that our lives are already so pleasureless, and that time is so precious, because life is so short.
- Most wealthy actors, game show hosts and contestants, newscasters, musicians, and teachers, must get D2BW teleprompting to make television shows, movies, concerts, lectures, etc. look more professional; without people forgetting their lines or saying unscripted things. Since D2BW teleprompting exists and excluded people

cannot tell the difference, it seems obvious that it would be used all over the Earth.

The Hutu and Tutsi, "second class" citizen nature of D2B included and excluded

Why is it so hard to get a date or make friends? Why do I never get the job? Why are so many people so rude and mean to me? One reason is because not only do those who get D2B not want to be involved with a person who does not get D2B, but because they are paid to discriminate and say rude things. There are a variety of reasons why D2B included avoid excluded:

1. Too difficult to communicate – they have to repeat their thought audio and try to describe their thought-images out loud.
2. Excluded are often tricked by remote neuron writing. The excluded can't possibly know a lot of important information that the D2B consumers know and so a D2B excluded seems dangerous to a D2B consumer. To a human who gets D2B, the excluded, who have never even heard of D2B, are like loaded guns or time-bombs walking around – the sophisticated computers of the neuron know exactly how to play them into a way that is violent or unpleasant to other people. To make matters worse, there is absolutely no way for the D2B consumer to give the excluded even a single warning like "Look out! That image was remotely sent to your brain…and the neuron writers are trying to trick you!". The D2B consumer must just remain silent and watch, and they may have seen, over the years, much of the abuse and violence done to excluded people by those with remote neuron writers.
3. In terms of a relationship it's like a person that can see and hear dating a person that is

deaf and blind. Perhaps these "Ice Castles" stories happen, but it is probably extremely rare. Maybe that "blind and deaf" excluded person is really attractive, but a serious relationship would be too much of a headache for somebody who can see and hear. Mostly this is the case for males, because pretty females are probably quickly included by wealthy included males. There are many better money and career offers, and simply many more choices for those who get D2BW.

4. Being nice to an excluded, makes the D2B consumer look like an outsider. This phenomenon is a classic throughout history. Nobody will defend the accused heretic, because they don't want to be accused of heresy. Other classic examples are: a person defending someone accused of witchcraft, a German person under the Nazis who was sympathetic to a Jewish or homosexual person, a white person friendly to a non-white person during segregation, etc. It simply looks better if you don't have any excluded friends that might call you or somehow intrude in your life and the lives of your D2B friends. All the basics of segregation completely apply.
5. Talking for extended periods of time with an excluded, who knows that they are excluded, and who talks openly about D2B – as is the case for me – makes the D2B consumer attract labels of "rat" from other D2B consumers and their D2B dealer.

The preferential hiring of D2B consumers over excluded is a massive and terrible truth. Many companies must only hire D2B consumers exclusively – there is not a single excluded person on their payroll. Imagine receiving D2B windows, seeing and hearing the thoughts of everybody around you – of course, you would not want to hire somebody who not only can't hear your thoughts, but is being actively manipulated by very violent and evil people all the time through remote neuron writing – and there is nothing you are allowed to say to the excluded to help to stop it- the D2B consumers can absolutely not help the excluded whatsoever or risk losing their D2B service potentially for the rest of their lives, or worse. The situation for D2B consumers is analogous to the "sunder commando" prisoners of the Nazi death camps, who could not tell each new group of prisoners that they would soon be killed with poisonous gas. In particular, in small businesses, where every penny counts, you don't want the risk of being a humanitarian and hiring a person with only 2 windows over their head and none in front of their eyes (a D2B excluded), because any mistake might end your business.

The excluded are the puppets of the included. Excluded people are used as pawns to sexually arouse D2B consumers. Trying to get excluded to do sexual inappropriate things generates a lot of money for the phone company and neuron owners. Simply put, D2B consumers pay the phone company a lot of money to see sex and violent videos, and this is a big incentive for the phone company to create sexy and violent videos – and many excluded fill the role of actors and actresses because the remote neuron writer never has to face their remote excluded victims or pay for their involvement in remote neuron written crimes.

Certainly one way to stop excluded from doing violent and sex crimes is to show and tell them about remote neuron reading and writing, dust-sized cameras, and direct-to-brain windows. Knowing that some violent or sex criminal might be writing to their neurons might make them think twice about acting on a remotely written violent or sex crime suggestion. Even simply letting excluded people know that their every activity is recorded with microscopic cameras by the governments and telecoms, and can be seen by many thousands of people, would help tremendously to stop many sex and violent crimes. In particular, I think excluded would not bite on illegal and unethical suggestions when they know that the captured images and sounds of them, although currently still secret, might someday be used as evidence to punish or jail them for the crime in the future.

In terms of the Hutu and Tutsi analogy, generally, using remote neuron writing, excluded males are feminized, and excluded females are masculinized. The movie title "Tootsie" is probably a double meaning play on the Hutu and Tutsi apartheid of Africa, in this case the "Tutsi" are the excluded whose males are feminized, for example with remotely written suggestive images of them wearing women's clothing. Other excluded males around them are sent similar suggestions to arouse them into potential male-male encounters, all through remote neuron writing and to the great pleasure and arousal of the Hutu D2B included males and females. Making your honest enemy extinct, by denial of D2B service, in this sense, is very erotic for the D2B owners and consumers. For an excluded male, images of them wearing a bra, women's underwear, touching another male, etc. are constantly beamed onto their thought-screen, in addition to activating neurons that cause the feeling of sexual arousal. For an excluded female, images of them wearing a penis or men's clothing, touching another woman, etc. are constantly written to their neurons, probably like the male, while pulsing the neurons that make humans feel sexual arousal. Excluded people, already deprived of and desperate for physical affection, simply view these thoughts as natural, not as being sent by other people, and are much more receptive to acting out the remotely written suggestive images. This fits well with the D2B desires. The neuron uses the beam to make the less interesting, less profitable, and possibly reproductive heterosexual encounters impossible for excluded.

One very interesting truth is that D2BW segregation is apparently a *global* segregation. For example, a person who gets D2BW in one country gets D2BW in other countries too. For example, when a person who is included to D2B in China, India, Russia, South America, etc. moves to the USA, they still receive D2BW, just like a person who is denied D2BW in the USA, that then moves to any other nation is still denied D2BW. One person I work with started receiving D2BW in India, and another in China, before moving to the USA (where they still receive regular D2BW). So D2BW is a *planetary segregation,* not a segregation that is local to each nation. A "D2BW consumer" gets D2B, preferential hiring, and other benefits in every nation, but yet there are no apparent logical criteria for who gets to enter this elite class and who is excluded. Equally shocking is that such a massive segregation has not been made public for, apparently, *centuries*. Trillions upon trillions of humans have kept the secret of D2BW from being made public for centuries. But integration, and the free-flow of thought-images and sounds in public, seems like the inevitable and hopeful future.

This is why, although it feels embarrassing, it is good to talk openly about the D2BW segregation, because your own right to reproduce and make a family is being denied, in addition to many other basic rights.

The "Ideological Genocide" aspect of the Neuron lie

Denying millions of people the ability to see, hear, and send images and sounds using thought is a massive and extremely effective breeding program that seeks to make the honest, poor, non-white, non-religious, and other groups of people extinct. The lie breeds true. Clearly those who lie have succeeded for many years, not only those who lie and suppress the truth about remote neuron reading and writing, but of course, over an even longer period of time, those who continue the lies and mistaken beliefs of the religions. Those who have no trouble lying about remote neuron reading and writing are readily included, get lots of sex, and reproduce making lots of offspring, while those who tell the truth- about D2B, 9/11, the JFK and RFK murderers, or even the truth about science and religions have to live lives of celibacy and abstinence, not by choice of course, but because it is very difficult to find a mate when you can't see or hear thoughts, and remote neuron writing is in the hands of a bunch of murderers, assaulters, molesters and liars.

First, getting sex for D2B consumers is extremely easy. Only excluded have to masturbate. In fact getting sex is so easy for a D2B consumer, that it becomes boring, and many of them have to abuse excluded people, to keep the D2B male erect, and to make their sex more stimulating. The D2B consumers live out their sexual fantasies by inflicting their fantasies to be acted out by excluded through remote neuron suggestions, of which perhaps 70% are acted on by excluded, while only perhaps 15% of remote neuron written suggestions are acted on by D2B included or excluded who know about D2B writing. Once you know about the existence of remote neuron writing, and this massive "denial-of-service genocide" and "extermination" through hording of neuron reading and writing, that you are on a stage with many voyeurs watching, you learn quickly to reject the suggestions of the neuron writers and focus on trying to get a date, to make a family, etc. But even so, with remote neuron writing in such evil hands, an excluded person even getting a kiss with an included or with another excluded must be extremely rare.

Many times I am sent the thought-audio of a woodpecker pecking, and I think what this means is that D2B women are constantly bombarded with D2B video invitations and offers from D2B guys, some offers, are probably for lots of money for little work in return. Of course, males who are being denied D2BW can’t peck on any females and so lose out on any chance of getting a date, a kiss, and ultimately to reproduce. D2B females probably immediately adapt to the money for pleasure system, and/or the pecks of the many D2B males; an excluded male, no matter how handsome, smart, and talented doesn’t stand a chance against the many males that can use D2B to communicate with both included and excluded females.

Making excluded live lives of celibacy and chastity is erotic to D2B consumers in the same way that they desire veal. Seeing a sexually frustrated celibate never-touched male or female is highly erotic to D2B consumers. Much of D2B consumer sex is built around teasing excluded people by flaunting the chance of sex, but absolutely denying them any chance of even a kiss. The excluded have to masturbate, but the sex cup overspills for the D2B consumers. Where the excluded are starved for even the slightest touch, the included, who see nude people and their thought screens all day, and can get all the sex they want, have to come up with new creative ideas to make the nude human look sexy, and to make sex seem exciting. As I said, non-violent extermination of your honest and lawful enemies by denial of service and imprisonment is actually very erotic for many people, in particular wealthy neuron owners- maximizing their physical pleasure and minimizing the pleasure of the opposite side. Imagine how easy it is to simply remotely give a human, in particular a D2B excluded, a "bad feeling" about another excluded. It seems likely that computers are programmed to automatically remotely neuron write to stop any kind of romance between excluded people. Misleading excluded people with remote neuron writing is very easy. It's hard to imagine an excluded person having the neurological strength to reject the natural inclination and memories flooding their thought sound and screen.

Not even warning the excluded about remote neuron reading and writing is all part of the massive "ideological genocide", or "excludocide", and is one of the most shocking and inhuman aspects of the D2BW "invisible" segregation.

D2B discrimination goes beyond gender and race; being D2B included or excluded is much more important than gender or race. D2B consumers generally don't hire excluded people, only excluded people hire other excluded people, so, as a result, excluding a person is to extremely reduce that person's chance of survival, because it lowers their chance of being hired, which lowers their chance of gaining income to buy food, shelter and the things they need to survive and reproduce.

No person who gets D2B is going to reproduce, date, or even befriend a person that does not get D2B. It reveals an unpleasant truth about D2B relationships: that the relationship would not exist without both people getting D2B service, and would probably fall apart if one was suddenly excluded- what a deep relationship eh? Losing D2B is probably analogous to one mate suddenly becoming deaf or blind. The only chance an excluded has to reproduce is to find another excluded, and even then, remote neuron writing is definitely used to cause chaos, and to lower the chances of success.

Hand-held and even microscopic beam weapons were probably first invented centuries ago

It sounds crazy, but yet if many people have been seeing and hearing thought for centuries, certainly there must be portable and microscopic particle weapons that can burn people. This includes the very dangerous hand-held maser (like those on "Star Trek" and "Star Wars" but invisible to animals). Many people are not even aware that a CO_2 laser, which is public information, can already cut through metal, and so certainly can cut through muscle and bone. The myth is that a maser (and laser) of high power requires a large device, but it seems likely that not only a hand-held device that can cut a person in milliseconds must have existed centuries ago, but that even microscopic devices may be able to seriously burn and cut living objects. I have felt tiny pin-point burning feelings on my body before and perhaps these are from microscopic laser devices. These devices may be used by two large sides to keep each other in "check"- one side attacking until the burning delivered in response is too intense. Imagine the strongest possible handheld lasers in the hands of average people, for example an old person accidentally slicing off their foot or their neighbor's hand in an instant. But believe it or not, the hand-held laser, like the ballistic gun, is mostly obsolete, surpassed by the far faster, computer controlled, microscopic particle devices which are always armed and pointed at most people every second of our lives.

The importance of "particle supremacy"

"Particle supremacy" is used to hold a monopoly on particle technology. Simply put, if somebody, a brilliant excluded, or even brilliant D2B consumer, tries to duplicate remote neuron reading and/or writing, all of the already existing nano-devices and sophisticated particle device infrastructure can be used to easily disrupt and destroy that effort without any chance of detection. Beyond that, all publishing is highly controlled by the all-powerful neuron owners, and since all major companies are owned by D2B consumer "hostages", they all must bow to the neuron owners, as strange as that may sound.

Without a doubt, micro and nano sized particle weapons are the top of the food chain of most dangerous and powerful weapons. Tanks, bullet guns, robots, missiles, macro air ships, are all mostly obsolete as weapons, and not worth all the tax money spent on them, because they are all penetrable by tiny particle devices. Larger ballistic guns could possibly have value in breaking through thick barriers for the side that has particle supremacy. Any next "World War" 3, as may have been the case for the other two World Wars, might be fought and decisively won in milliseconds. The first and most important war is the nano and microwar, since those devices can easily destroy any larger devices, and are the only devices that can stop each other. But it's not clear, and I am excluded, so I can't be sure.

The light particle, or "photon" is the most dangerous projectile, and particle weapons are easily the most dangerous and deadly weapon known on Earth, and perhaps even in the Universe. Light particles move much faster than a bullet can, and can be sent invisibly from tiny microscopic devices.

Remote controlled microscopic particle devices created more opportunity for wealthy violent people to do remote violent crime and never be caught. The remote nanometer scale particle device is almost the perfect murder weapon, because it is 1) invisible: nobody can see a device firing light particles with x-ray and other invisible frequencies. Microscopic particle devices are very difficult weapons to destroy, because you have to pin-point the source of the particles, which could be a variety of tiny dust-sized hovering devices. 2) Leaves no trace: simply holding a lung muscle so a person cannot breathe for about 3 minutes is enough to remotely and invisibly kill a human or any other species with a lung, and all any coroner or loved one can say is "death by natural or unknown cause"; there is no wound or bruising. 3) Can be remotely controlled: the murderer, assaulter or molester doesn't need to be physically near the victim at all. Since the dust-sized particle devices form a chain from attacker to victim, a person in some underground bunker on one side of the planet can remotely make the nose itch of another person on the other side. Remote controlled nano particle weapons opened up new opportunities to remotely torture people. This is used on me, and I'm sure on many of you, all the time – my muscles are moved, many times, a right finger will be moved, just as I am falling asleep to do "sleep deprivation", your muscles are moved to make you "accidentally" drop things (many times into a toilet or onto a bathroom floor), to "accidentally" cut a finger when cutting food, to "accidentally" burn your finger or arm on an oven grill when putting in or removing food, given a headache, or given a remote "mystery" skin burn or cut.

Throughout history many murderers have adopted a "right" or "left" theme to create conformity and unthinking obedience. Judging from the many murders that we *do know about* done by the current "conservatives" (in the United States, the

"Republicans") of many liberals and popular pro-democracy leaders like the Kennedy's, MLK, John Lennon, and mass murders like 9/11/01, 7/7/05, Utøya Norway, etc., this modern group of evil, like the Nazis, has flourished using the "right" theme to infect the planet with their violence and destruction like a deadly virus. They infect the governments and major media, and just like a violent virus, harness poor, and otherwise nonviolent people in the army and the public to kill for them with very little choice and without knowing anything about RNRAW and D2BW. There have been terrible un-democratic rogue murderers on the non-religious so-called "left", like LBJ, Stalin and Pol-Pot, and there are violent rogues on both sides. Historically, the conservative "right" has probably produced the most first strike violent leaders like Hitler and the Bushes; people who harness the power of the Church, and continue the worst of the Church's brutal and barbaric traditions of harshly torturing and murdering the very best and brightest minds of Earth for petty and trivial reasons and worse- for telling the truth. Clearly they are not right or left, but simply the "first strike violent" and "big liars". A battle of "sides" or "colors" will, I think, in the long term, always lose to a battle of "ideals". Battles of "true versus false", "fair versus unfair", "stop violence versus start violence", etc. will always win over any battle of simply "left versus right", "blue versus red", "north versus south", "east versus west", etc.

Most if not all D2B owners, and many D2B consumers, do not just sit back and deny millions the ability to see, hear and communicate by thought, but they are funding and undertaking a massive and very active war against truth and against those telling the truth (in particular the truth about remote neuron reading and writing, light as a material particle and the basis of all matter, the expanding universe and relativity theories being wrong, and the truth about violent murders like those done by Frank Fiorini, by Thane Cesar, by the controlled demolitions on 9/11/01, etc.). They don't just sit back and allow free speech and the free flow of information to take it's course, but instead wage a constant and often violent war against the truth and against educating and informing the public.

Methods of remotely killing people with remote controlled microscopic flying particle devices

1) Remotely hold a critical muscle contracted, or prevent it from contracting.
2) Induce cancer (William Rollins demonstrated this on guinea pigs back in the 1800s)
3) Use remote muscle neurons to induce a fall down stairs, off a ledge, car accident- how many car accidents were actually remote murder?
4) Use "voice of God" (neuron writing sounds to thought-audio and ear-audio) to trick excluded into suicide, homicide, or assault. Another technique is to get them to do something illegal, like a sex crime (excluded can't get sex, and so this is an easy technique), get them caught, and then remotely murder them in prison which is easier.

Remote neuron writing is used to help reinforce false stereotype

One use of remote neuron writing is clear, and that is to reinforce false stereotypes that the D2B owners want to trick more people into believing. The excuses given for excluding people are usually the same excuses, and they center on "pervert" and "crazy" (notice how "violent" doesn't seem to be a big issue- but it can be if it's linked to sex). Most of the time, these labels are applied to people who are just as normal and sexually aroused as any average person. But like the 9/11 19 hijacker story, the Oswald story, the Jesus rose from the dead, and the Moses parted the Red Sea stories – if a lie is told enough, many people will believe it and any opposition to the lie is viewed as unorthodox, then any person who supports the truth is viewed as a black sheep. All of remote neuron writing is used to try and reinforce these myths; for pervert: 1) by sending sexual suggestions to excluded, 2) by literally moving their eyes to people's buttocks, breasts, etc., which D2B consumers can see on the excluded person's "eyes" screen, 3) by the nature of being excluded: an excluded has to look if they want to see a good looking person (and potential mate) while an included can stealthily just call up a live-image window of the hottie in front of their eyes without appearing to stare- and when I say a window of the hottie, I mean one that shows everything – what they look like nude, all the sex they've ever had, all of their most private thought images, their religious and political views, etc. The neuron owners and developers have taken this natural voyeuristic desire to "people watch", to see people in the nude and having sex, and to hear their conversations, what is traditionally called "perversion", to a surprisingly advanced state.

Labels of "crazy" tend many times to be a cover-story or excuse to try to discredit and defame those telling the truth. Classic examples are those who express doubts about the official major media (television and newspaper company) big lies: that Oswald killed JFK, that Sirhan killed RFK, that 9/11 was not a controlled demolition, that people haven't kept RNRAW secret for centuries, etc. who are quickly dismissed and defamed with abstract words like "wackos", "kooks" and "cranks". Also traditionally, those who express doubts about the ridiculous claims of the religions are often, ironically, labeled and viewed as crazy. In general, people who lie are rewarded, while those who tell the truth, about remote neuron writing, 9/11, religions, etc. are demonized and defamed. Scoundrels and mass murderers like Bush and Cheney are labeled heroic, while the actual heroes, like those telling the truth about 9/11, are labeled crazy. So the non-religious and honest are the first to attract the label of "crazy". Remote neuron writing is then used to reinforce this stereotype.

First, many of the claimed psychiatric "disorders" are not based on clear, diagnostic, and physiological science, and many times the prescribed "cures" (massive involuntary drugging with experimental drugs, electrocution, and physical restriction) are radical, brutal, and many times illegal responses to petty or misdemeanor nonviolent behavior. Secondly, because of remote neuron writing, many people don't realize that what is being called a psychiatric disorder may actually be more accurately described as "remote neuron abuse of a direct-to-brain-windows excluded". For example, what is being called "manic depression", "neurosis", "psychosis", "paranoia", "unusual behavior" – may in fact have an external source. An excluded is very easy to remotely control since they have absolutely no knowledge of remote neuron reading or writing. They may follow the "voice of God" in their head. Many people might not even be aware that many victims and prisoners of the psychiatric system are honestly and accurately describing remote neuron writing when they state that they "hear voices in their head". Many people recognize the classic claim of "people hear my thoughts" and "people watch me in my house". How vicious are the D2B consumers who know that what all these "crazy patients" are saying is, in large part, true, and is the result of the massive secret of remote neuron reading and writing, but still say nothing, and allow drugging, and other "treatments" to occur. Other classic remote neuron example "symptoms" are: an excluded person mistakes someone's identity because through remote neuron writing an image was remotely overlaid over the face of a stranger for a millisecond- the excluded may run up and say "Fred!...oh wait...I'm sorry...I thought you were my friend Fred". With remote neuron writing an enemy of a D2B consumer can be made to have an unnatural devotion to a celebrity, or somebody they have never met, or a person that abuses them. Movie and television owners, and individual wealthy people may buy remote neuron writing on to D2B excluded (who have never heard of it), for example, to bulk write images, or send audio samples, and remotely evoke specific memories to maximize the attention directed at the celebrity they are trying to promote, and therefore to increase the money they receive. Even the enemies of popular people may remotely send images and sounds to try to scare and bother the popular person by constantly bombarding the thoughts of excluded people with memories, images, and sounds of, or relating, to the popular person to unnaturally create a strong devotion, or alternatively a negative impression in the mind of the excluded person. Another technique is to periodically send the excluded an image of a neighbor or coworker. Then the excluded focus their anger on that neighbor or coworker, and integrate those people in their thoughts into elaborate inaccurate theories. In the absence of any kind of established and public scientific explanation of remote neuron reading and writing, some excluded form theories of their own, to explain how many, or a few particular people around them, can apparently see them in their house and hear their thoughts. For example some may think that their metal fillings allow people to remotely control them, that there are tiny "peep-holes" in the walls of their house, or other inaccurate explanations. Many of the explanations center around the belief that God is communicating with them. Thinking that God is telling you to do something, no matter how embarrassing, violent, or illegal, will cause many excluded to act on unusual suggestions. Many excluded want to show their absolute allegiance to God and so obey the remotely neuron written suggestions. Excluded may go nude in public, or even commit gruesome acts of violence, in order to fulfill the wishes of the supposed God who communicates with them through their thoughts.

One popular technique is to play on sexual jealously in monogamous minded excluded. Excluded are routinely caused to experience suspicions and sent sounds and images that suggest that their mate is having sex with other people. Playing on sexual jealousy is a major part of the criminal remote neuron reading and writing empire. It is one of the easiest ways to make a male (or even a female) violent, and to destroy a relationship. Remotely writing images and sounds that try to trigger sexual jealously is probably one of the top ten most successful methods of getting excluded people to do violence, and videos of violence increase the income for the telecoms which sell the videos to D2B consumers. So watch out for remote neuron written attempts to make you sexually jealous- it's better to keep a cool head and realize that people need to be allowed to have sex with whomever they want to consensually, and that much of what is neuron written are lies. Another technique of trying to create sexual jealousy is to make an excluded think they have a chance with some D2B consumer, then paying the D2B consumer to reject the excluded, and walk past them with some D2B mate. To "tease" the excluded is very erotic to many D2B owners and consumers, but also serves the purpose of reinforcing the addiction and obedience of D2B consumers to the demands of the D2B owners; they know to obey the neuron or else they very easily could become that poor celibate excluded "drone bee". A D2B couple may make themselves more aroused by walking by (individually or together) excluded and then paying AT&T to see the excluded masturbate to thought-images of the D2B consumer to enhance their sex (or they might be paid by wealthy D2B consumers to do the tease). One D2B consumer kindly tipped me off to this saying "they're going out", with the helpful double meaning of "that person has a boyfriend already, don't waste your time", and of how some D2B consumers figuratively and no doubt literally "urinate" on excluded people with this "walk-past" teasing. One positive aspect of this barbaric practice is that the D2B consumer probably finds the excluded person attractive. One good way for an excluded to defuse the tease, is to approach the teaser and talk to them for a few seconds, and perhaps try to get a date, because the excluded will then recognize quickly if the D2B consumer is friendly and might actually date them, or if they have the rude, brutal, Nazi-like elitist "leave me alone", poop on the subhuman excluded, viewpoint so typical of many D2B consumers who get anything they want delivered instantly right to their eyes.

Many excluded may express anger or become angry often because of remote neuron writing. Is it any wonder that an excluded person would have anger directed at direct-to-brain consumers around them who are shilling, that is, getting money to say rude comments like "psych", "gay", "vert", etc.? Excluded simply have not the vaguest idea that the D2B consumer people around them are saying these rude things for money, and that the sound and/or image in their mind is not from God, but is probably from wealthy uneducated and violent people who have a shockingly extreme technological advantage of being able to see, hear and write to their thoughts. The poor excluded never even know that such a thing as remote neuron reading and writing is not only possible, but is highly developed, and is in constant use, and mostly constant abuse. Imagine the so-called psychiatric delusion and problems that will be suddenly "cured" when direct-to-brain windows goes public and everybody can see their thought-images of the past.

Another stereotype that remote neuron writing is used to reinforce is the myth that non-white people, by the nature of their race, are more likely to commit crime. By remotely writing suggestions on the neurons of non-white excluded people to do crimes, remote neuron writing is used to trick them into committing crimes. Then the news stories are filled with stories of crimes where non-white excluded people acted on the criminal suggestions and committed crimes, which reinforces this inaccurate belief.

For many years bad people have been the most powerful on Earth and will continue to be for some time into the future

Clearly evil violent dishonest people are at the top and have been for a long time- certainly since the rise of Christianity in the 300s but probably even before then. That is the only way for the murderers of the Inquisition, of the Kennedys, the WTC 9/11 victims, etc. to go unpunished and free, in addition to the many neuron writers who have murdered many men, women, and children, like those kids recently murdered in Norway. The simple fact that direct-to-brain windows has been a secret for an absurdly long time, of the shocking disregard for science (only a tiny minority supporting a theory as obvious as evolution and knowing very little about the history of science), and the shockingly long duration of the belief that Jesus and Mohamed had supernatural powers, etc., are all evidence that, for a long time on Earth, bad has been on top, in control, and good has been on the bottom. Whoever shapes and controls much of the planet, the most wealthy, clearly have a violent and dishonest majority view– people are murdered, and their killers are never even identified, massive lies like the expanding universe and 9/11 19-hijacker story continue to be believed for decades, millions are denied even knowing about seeing and hearing thought, let alone regularly receiving it, etc. – just terrible values.

Everything runs on money

One good way of looking at life on Earth is that absolutely everything runs on money. Almost everything that is done, is done for money. With the development of remote neuron reading and writing, and direct-to-brain windows, came an extremely rapid free market. Now, with D2B, many people can receive and accept offers for money directly in front of their eyes. Money is the fuel of people, and many people will only do something for money. But the myth is that famous people are famous because they have special talents. Some amount of skill and popularity certainly plays a role, in particular in sports, but much of the success depends purely on money. For example, if you want to be on television or radio, you have to pay the owners for that time, which is very expensive – but this transaction is, apparently, all done secretly through D2BW. Just like this book – I have to pay to have it printed, and pay to advertise it.

There are entire secret industries that arose with the invention of D2BW. The "shill" is an industry where D2B consumers receive thousands of dollars to say annoying, upsetting, and misleading things to those excluded from D2B around them. And that shill industry extends to D2B consumers – people can pay to give them a bad dream, and annoying D2B memories, muscle moves, sounds, images, etc. Newspapers, radio and television news are perhaps the most misleading – Fox news, is a fine example, but they all exist strictly from payments of money to deliver information that the buyer gives them. For Fox news it's basically the 9/11 murderer group who fund so much of what is told to the public as if it was actual news. Notice, for example, how no major news source ever reported the discovery of D2B, the truth about Frank Fiorini, Thane Cesar, 9/11, 7/7, etc., or even that anybody has any doubts of the

Republican and Democratic representative and major media company's version, or even that there *is* a second version or explanation. Mostly, the major media are advertising companies – and the most convincing ads are disguised as "news"- all of which is paid for and purchased by D2B consumers and owners through the D2B silent and invisible (at least to excluded) interface. Think of the D2B consumers who turn down thousands of dollars to shill and do other unpleasant things. It must be very difficult to not accept an offer for $5,000 just to say "gay", or $500 just to walk by some excluded person and cover your mouth with your hand. Perhaps the view of those D2B consumers who pass on all the money is that there is a higher potential future money for not shilling or taking money to do immoral or unethical things.

It is simple to understand that all the television, radio and news companies survive from large payments for advertising, and I think when D2B goes public, the public will finally see that probably every single image, sound and sentence broadcast and published by the big media companies is a paid-for ad; many times paid for secretly, using the "absolutely no-paper-copy" D2B network. This is obvious when you realize that, even just for here in the United States, all of these terribly obvious violent murderers are on the loose- those who did the 9/11 controlled demolition, the killers of JFK, MLK, RFK etc., but the major stories being funded, printed and broadcast to millions of excluded people are mostly about trivial non-violent activities. It's clear that one goal of those who fund and own these companies is to change the focus of the excluded people from serious violent crimes to a more anti-pleasure, anti-drug, anti-free-info focus. For example, in the interest of survival, the clear focus in my mind is to show and tell the truth about remote neuron reading and writing, evolution, the history of science, and our future as a star cluster, about full democracy, total free info, violence as the big evil, pleasure as something to consensually teach, and openly celebrate, etc. In particular a clear important focus in my mind is that the public should see all the crimes of the worst of the violent, to remove the worst of the violent from remote neuron writing, and from being able to do more violence. But showing the truth about 9/11, the Kennedy killings, the galvanizations, etc. and exposing those murderers who live free and unseen on the loose is not the main story, instead, accusative stories without any actual physical evidence about inappropriate non-violent sexual activity (in particular with children) are the main stories. To their credit there are stories about poor (presumably D2B excluded) people who do violent crimes, with low-tech fists, knives and guns, etc. but then always the explanation is not remote neuron writing puppetry, but the immediate application of unlikely psychiatric theories. Even with the remote neuron writing "television" (the unconsensual one on your thought-screen and in your thought-audio that you can't turn off or block), the views paid-for and written are always that sex and sanity are the focus and topic of everything; stopping violence and exposing widely believed lies is never a concern. The "neuron" is so powerful, in particular with their remote neuron writing on the "sensitive" brains of excluded people, that there is a strong argument that we can read any news story by replacing the names of the people with "neuron", for example, we can replace "the suspect did this" with "the Neuron did this". When a person murders, we must ask "why did the Neuron murder?", when something is stolen "why did the Neuron want this stolen?", etc.

There is probably an unavoidable and somewhat brutal natural economic and biological truth in the deciding by the majority of D2B owners and consumers which humans may continue to live and which may die. It may even be that the main reason some person is still alive is because they are doing something (telling people about RNRAW, 9/11, helping, etc.) or have some kind of value (are pretty, funny, etc.) that justifies the cost (in money and even perhaps in lives) to protect them from the many violent people on the other side who remotely end the lives of many people (who, it may be, do not justify the cost to protect them) all the time.

There is a funny video from SCTV, "Dan Money"[3], where Joe Flaherty has to pay everybody to cooperate, for example, the judge asks the jury foreman "has the jury reached it's verdict?" to which the foreman says "well your honor, maybe we have and maybe we haven't...", and Dan Money has to pay the foreman to get the verdict he wants. We can laugh at this funny 1970s television cop show parody, but this skit may be hinting at a darker meaning given what we know about the long-lived secret of D2BW. The truth is that millions of poor D2B consumers march all over with little money windows in their eyes. It seems very likely that direct-to-eyes money offers through D2BW certainly have corrupted the justice and many other systems. Probably many people are called to court or jailed because some people (including members of police, a jury, and/or a judge) were paid money through D2BW.

[3] "Dan Money", SCTV, 11/28/1977
http://www.youtube.com/watch?v=jvEGOEA0OiA

Violence and sex shows and games

It's gruesome to talk about, but it seems likely that there may be pre-murder "shows" or "videos" that D2B consumers can buy from the telecoms. These videos may have narrators or commentators describing the potential murderer, the thought-images of the murder that the potential murderer is planning, the profile of the potential murderer and the possible victims. These videos are probably advertised in pop-up windows to the D2B consumers who must pay to see them. Just like there are "get money" windows, of course there are probably many more "pay money" windows. In addition to the pre-murder videos, there most certainly must also be "post-murder" videos of the murder during and after the murder from the eyes of the murderers and victims, and from micro and nano-cameras hovering around the scene, which the Neuron companies sell to the public. These post-murder D2BW videos are probably much more common than pre-murder D2BW videos, because there is no chance of D2B consumers interfering with the murder after it occurred.

In addition there may be sounds of thought-audio cheering, and applauding. When a violent scum is murdered good people probably applaud in their minds, and when a decent person is murdered the criminals applaud in their mind. A classic example is the way US President George Bush jr. clapped at the same time a plane crashed into the World Trade Center on 9/11/2001. Bush sees this plane crashing into the WTC in his D2BW and then he and other like-minded D2B consumers must clap to reaffirm approval of this radical direction that they took.

In addition to the violent videos there must be a big industry of sex videos, and the sex videos of the neuron must be very different from those available to the general excluded public. For example, the D2B videos certainly are of real-life sexual events, many of which are not scripted. Many of the D2B sex videos certainly contain violent crimes. Without doubt many of the D2B sex videos contain people under the age of 18. Probably the videos of excluded people who bite on sexual suggestions are very erotic – in particular because they are so innocent in not knowing anything about D2B, are very sexually frustrated, and don't realize that many people are watching them. But even the videos with D2B consumers may be quite erotic because they are probably much more sexually experienced than excluded people. In addition, D2B consumers can participate in the massive and fast moving money for pleasure market in the D2B, which the excluded cannot access.

In terms of games, there may be gruesome games that revolve around sex and around violence. I hesitated to describe this because it could be unpleasant to some people, but in the interest of knowing the truth about what D2B consumers and owners might be doing, I decided to include it. I can imagine that there may be games like "which one lives?" in which viewers vote to determine which of two humans will be allowed to live. Most of these games probably involve excluded people who are like "throw away" people, and the excluded certainly can't see to complain. Probably "Russian roulette" is played with various death-devices like guns, perhaps with prisoners of war, or with people who agree to take the suicidal gamble for thousands of dollars. Some other gross games might be "snuff the excluded"- trying to use an excluded or D2B consumer with a gun or weapon to hurt and/or kill another excluded (or even a D2B consumer) using remote neuron suggestions. You can see that something like that was probably done to John Lennon, Gabby Giffords, and many other people. Perhaps wealthy people may use their vast wealth to pay for "irony" sex to occur at the same time they are trying to kill excluded or D2B consumers (for example some poor educated young woman who is gang raped, or takes thousands of dollars to "hand" a bunch of neuron criminals while they try to remotely kill her husband, boyfriend, or a popular enemy of the funders). Possibly the phrase "bang bang" which I just heard in thought-audio, sometimes applies to those kinds of videos. We can only imagine what kind of vicious things the ultra-wealthy D2B owners participate in to make them aroused and entertained.

The sexual aspects of the direct-to-brain segregation

One thing is clear and that is that D2B consumers can easily get sex, and those excluded from D2B mostly cannot get even a kiss, and certainly not sex. I apologize if you are a person who is offended by talk about sex. I prefer to take a scientific view of sex as a natural biological phenomenon on Earth. It is helpful, in the interest of knowing the truth and protecting ourselves, to understand that much of what is happening around us may involve some kind of secret hidden sex. Beyond that, consensual sex and talking about sex is nonviolent, and we need to educate ourselves about the science of sex, anatomy, and physiology.

How sex operates for D2B owners and consumers is not entirely clear and we can only guess about what they do. Clearly many D2B consumers hint about "coatings" and "coverings" and "huddles" which implies that a number of males coat or cover a single female with their sperm. When you can see and hear thought, it is very easy to hook into the system of getting sex. With D2B, the D2B consumer can easily and quickly know who has a sexually transmittable disease (STD), and who doesn't – while excluded can only guess, and probably default to not wanting to take a chance of catching an STD. It may be that excluded and D2B consumers may even be targeted with potential sex partners who have STDs – 9/11 and so many remote murders shows that the neuron owners and many wealthy D2B consumers are vicious, cold-blooded murders and assaulters, so giving excluded enemies an STD would be mild in comparison.

The logic of why the D2B pleasure industry is probably huge is simple, you can either regularly masturbate yourself, or have somebody else masturbate you, you can either think of a woman or man, or actually touch one. Of course, most people prefer to actually touch. The various forms of love, like physical caressing, simply feel good to many humans. Beyond that, when you see many and most people participating in a money-for-pleasure and dates system – most new D2B consumers must happily and readily jump right in, and their lives suddenly become much happier and pleasure-filled. I don't know for sure, but it may be that for many D2B consumers, that males have to pay females for a kiss, a date, etc. has become more of a symbolic formality like tipping.

Clearly there is a massive money for physical pleasure system for D2B consumers. Getting money-box windows and videos direct-to-brain (appearing in front of your eyes as semi-transparent windows or on your thought-screen) is an extremely fast way to make transactions. The D2B consumer simply thinks the audio "yeah", "more", or "no", or even just draws the window closed in their mind, and the computer understands to close the window or "up" the offer.

Put yourself in the place of a poor female or male: they need to pay their rent, have no job, can't live with parents or relatives, and there are the money box windows with easy money – and all you need to do is to just let somebody touch you, or kiss, or rub somebody– for thousands of dollars – enough to pay your rent, for a car, or for a college loan, etc. Once robots are doing all manual labor tasks, one of the only remaining paying jobs for most humans may be physical pleasure for money, so it makes sense to officially and publicly end the prohibition on consensual prostitution for the D2B excluded people too.

One insider hinted that perhaps half of the D2B consumers do pleasure for money, the other half perhaps opting for a more traditional monogamous pleasure-based relationship. Clearly D2B consumers can browse the profiles, directly to their eyes, of many potential mates to make a family with, including seeing each potential mate in the nude, their lifetime accomplishments, and many intimate details of their lives. Excluded people only have the Internet and must choose from those in their immediate surroundings, many of whom are D2B consumers who will not even befriend them.

Some of the more well known aspects of D2B sex rituals are: "see you laid-her", "laid out", and "cuckoo". It seems possible that many D2B consumers can easily seduce excluded people, because they can see their thought-image, hear their thought-audio and deliver them exactly what they expect to hear. So many males, for example, may have sex with a large number of included and excluded females – they may "lay-out". The saddest part is that the excluded females think the insider guy is being monogamous, and this may arouse the D2B male even more- to be tricking or "using" the poor excluded female. The cuckoo is a bird that replaces an egg from the nest of some other species of bird with one of her own eggs, and then flies off. The other birds mistake the cuckoo egg for their own and raise the baby cuckoo. Many women get D2B, because the neuron owners know that women can generate a lot of money for them from the many wealthy direct-to-brain "johns", men who use Ma Bell, or "*Madam*" Bell to browse, shop, and eventually pay D2B women for physical pleasure services. Isn't it a surprise to find that AT&T is perhaps the biggest and most successful "pimp" of history? Much of the money AT&T gets is from the work done by many "hand-maiden" women. There is probably a similar analogy with the many women operators that work for the phone company. Direct-to-brain females must get many many money windows with offers for sexual jobs. It sounds probably very blunt and crude to excluded ears, because the well-oiled D2B pleasure-for-money industry has reduced the system to quick terms, but many transactions may be for "hand", "suck", or "hole". It must be very tempting to accept these easy, fast and STD-free jobs. Many jobs probably provide lots of money for very little work. For example $5,000 dollars just to masturbate a guy, $1,000 just to let some guy fondle their breasts for 10 minutes, or for a kiss, or $500 to caress their butt for 5 minutes. You can see how an excluded could never compete – they have no idea that you have to offer money, and that not only that, but lots of money. So a D2B female may find an excluded male who she can live with, marry, and perhaps even have kids with, while working during the day getting thousands of D2B dollars for D2B sex jobs, even just for "handing" D2B guys, while her excluded husband is at work, and the excluded husband never finds out about it, how could he? But certainly D2B women must get pregnant by D2B guys, and an excluded male raises the kids not even knowing that they are not from him, and so you see the analogy of the "cuckoo" (although with the D2B cuckoo, a male fertilizes the female egg and leaves the child to be raised by a different unknowing excluded male with the original mother).

Much of the "see you laid-her" (which D2B consumers hint about), is that wealthy D2B males will pay a female to walk by some excluded male, to "tease" him, to make him think he might get a date or maybe even sex. Sometimes they may even be paid to make sexual gestures to the excluded. Then they pay this female for sex while they watch the recorded thought-images that the male had of the female when he masturbated. For a female, seeing a male with thought-images of herself during his masturbation is probably a turn on for her. It probably is erotic for D2B consumer male(s) having sex with the D2B female(s) too. It's tough for an excluded not to think about the people around them when masturbating, because that is all they have to choose from, and excluded are starved for even the slightest hint of affection anyway. One D2B female audio I received said it well (paraphrasing): The excluded gets the trick, the D2B consumer gets the treat. The excluded gets the lie, the D2B consumer gets the lay. It stands to reason that since no D2B consumers ever respond to your thought audio calls of "come here, and let's talk", or "how about a kiss?", or interact openly with your many thought-audio questions and comments, that talking out loud to them would make no difference either. But perhaps the D2B "teaser" might actually "work out", or might respond, but only if you talk to them out loud – it can't be ruled out in theory. Plus, I don't think people should view "teasing" as negatively as perhaps many might, because, although it is a terribly imbalanced, unfair system with evil people in charge, still, it is likely that "the teasers" are sometimes attracted to the poor excluded target of the paid-for teasing. Being friendly to a person you might not necessarily have sex with is natural, and so in some sense, innocently, and without any mean intention, a person may appear to be teasing, but is only being friendly. For some people courtship takes a long time and there are many tests analogous to spider or swan movement-matching courtship. Some of teasing can be explained as a need for variety, which drives sex for many people. The need for variety in sex is similar to the need for variety in other parts of life; we don't want to eat the same food, watch the same movie, and listen to the same song every day, because it becomes boring after a while. Many times people may have a partner already, but are open to, or actively looking to find a new and better match. Beyond that, the teasing of excluded people is nonviolent; and of course, violence is the big evil, and the thing to fear and to be angry about.

Beyond watching an excluded person masturbate to thought-images of the paid-for "teaser" who walked by them, another aspect of "see you laid her" is that a D2B woman may be aroused by an excluded male (whose image is clearly seen on her thought-screen during masturbation and/or sex), and D2B males may then pay this D2B woman to remotely "abuse" the excluded guy which many D2B owners and consumers might pitch money in to see and may find erotic. There is a double-word score for the evil D2B in that, possibly, if the excluded male was included, he and the D2B woman might mate and make a family together; so they keep both (perhaps like-minded) people from reproducing with this scheme.

Another possible scheme is that a poor D2B woman may be paid a lot of money to have sex with the co-workers, worst enemies, or best friends of an excluded person who often has images of her on his masturbation screen, or who often appears on her masturbation screen. Even if a D2B woman doesn't take the money, the D2B consumers around excluded people are paid to shill words to that effect by implying that this kind of sex happens, to try to amplify the sexual frustration of the excluded.

Even young post-pubescent kids in high school are the subject of sexual remote neuron writing. The males the D2B helps are many times violent and uneducated. Remote writing is used to help these males get sex, and to prevent the opposition, their enemy, the bright, non-religious (or simply non-Christian) males from getting sex. No matter how handsome they may be, the smart and non-religious males then lose out – not naturally, but all through remote neuron writing with an extreme 9/11, Kennedy killing, Inquisition-like bias. As if by magic, the young females just mysteriously feel aroused by the crude guys, while the nice guys mysterious say all the wrong things, are mysteriously too shy, make all the wrong decisions, and just produce a negative impression in the young women's minds. This exact same, "sexual shaping" using remote neuron writing continues after school too. Part of this system is already openly clear in the choices of who gets D2BW- all the most violent and crudest of men, like the planners of 9/11 and the Kennedy murders. Simply receiving D2B increases the chance of sex and reproduction of the violent and crude males.

Apparently, the more frustrated the excluded feel, the more erotic it is for many ultra-wealthy, decadent D2B owners and consumers. While an excluded may be aroused by something as small as the scent or voice of a female, the D2B consumer can get all the sex they want, and so they have to create new and extreme situations to make sex exciting and interesting in order for them to naturally stay erect and aroused. For example, young attractive women may be paid thousands of dollars to "molest" excluded by remotely making them itch, or moving their muscle by following the commands they see in semi-transparent D2B windows, while some drooling thug holds them by a chain and has paid-for sex with them. I can only imagine what the public might see if D2B ever does go public.

"See you laid her" might also relate to poor women who want to see videos of people in their houses and their latest thoughts, but don't have any money, or even simply because the neuron forbids it, so in order "to see you", they have to do hands, sucks, and holes (ergo "laid 'er"). To see people they want to see, they have to do sexual favors, for example, have sex with the neuron owners or their friends. It is the classic story of "you don't get to see, until you…." and then fill in the blank.

In addition, in the process of a D2B consumer finding a mate, the D2B consumer may entertain the possibility of an excluded getting included at some time, and so send romantic thought-audio and images to the excluded, but all is for not if the excluded never gets included, which is probably the rule, and so it results in a system to the liking of the evil D2B owners and consumers who view such messages as a "tease", since probably 99% of the time, that friendly voice in an excluded head is from somebody they will not only never meet, but probably never even see, and who may not even exist.

Many excluded have very old fashioned sexual views, and so we have to remember that as long as no body is hurt, and there is consent, most sex is just a bunch of "goo"; violence is much worse. An important lesson for excluded people is that sexual jealously is stupid, because if you really care for somebody, you would let them have sex with whoever they want to- you would not try to control them.

It is unpleasant and shocking to write, and I have doubts about even including this in my book, but have decided that it is better to get as full an accounting of what may be going on as possible. It seems likely that there are also many violent videos that the neuron owners (again presumably the telecom companies) make and sell to D2B consumers, for example, "snuff" movies, where, mostly excluded people are murdered- some being murdered during some kind of sex event. Probably poor women are choked to death and the neuron owners sell the videos directly to people's eyes, not only of the scene, but from the view of the eyes of the victims and murderers, and of their thought-screens and thought-audio- which must be very unpleasant and shocking. This may be one aspect of how the D2B consumers, as some have said, apparently "a-pre-she-ate-it"; that is, seeing videos before a poor excluded or D2B consumer is murdered, to emphasize the importance of secrecy. The neuron owners know that D2B consumers pay hundreds of dollars to see such movies. It may be that many people who die were actually secretly murdered in this way. I can imagine a liberal victim getting thousands of dollars to masturbate a bunch of conservative murders, who may then assault, urinate on, and abuse the victim. There is probably an endless supply of poor and hungry humans who have no money or place to stay that do unpleasant things for money. Another classic scene might be that the D2B consumer thinks they are just going to do a "hand", but are then murdered- and nobody is punished because it's all a big secret, "under the neuron".

In summary, it seems likely that a very large and active money-for-pleasure market exists for D2B consumers, and that shockingly and idiotically, and perhaps erotically, they deny the public even an equivalent vastly reduced excluded service (since they work so fervently to keep even asking to be paid for or to pay somebody for pleasure illegal). It's tough to know without seeing the thought-screens of the wealthy and those owners of the telecom companies. Maybe we do live in a very sexually celibate age where even most D2B consumers wed, remain monogamous for life, and any thought or talk of "coverings" and "coatings" is only another big lie or futuristic wish by the D2B owners and consumers.

Common mistaken beliefs by people excluded from D2B

Being like a blind and deaf person, a D2B excluded has many common mistaken views. One mistaken belief is that you are going to tell the D2B consumers something they don't already know. This is a big mistaken belief that, of course, the neuron owners and consumers do nothing to stop. Perhaps there are exceptions, like the details of evolution and the history of science, the truth about religious history, the alternative views to the propaganda they receive, but generally, the D2B consumer already knows just about everything that you might possibly tell them- because, it seems likely that they have access to the thought-screens of many years of history – they already know that all matter is made of light particles, that light is a material particle, that the expanding universe theory is fraud, what you had for breakfast, what you just thought, your mother's maiden name, your social security number, all of your secret passwords and combinations, where you keep your valuables, what your most embarrassing moments have ever been – they know it all already as unbelievable as that is to accept. Talking out loud is basically obsolete for those who can see and hear thought, but may still have some kind of function, and must, of course, always remain legal and unpunished. Mostly, any time you see an image of some person who is a D2B consumer, and think of something you may want to say to that person – like rehearsing for a future conversation – for example thinking "when I see that person, I will say this..." – it's already too late- most D2B consumers apparently review the thought-images and sounds of any excluded people that thought of them (had their image in their thought-window or said their name in thought-audio), so they know long before any future conversation might happen what you plan on saying or doing. In short- you are having that conversation now, or they will probably be aware of that thought, long before you see them in person again. Most conversation for D2B consumers may be only in thought-images, being the fastest method of communicating some idea, but also perhaps in thought-sounds too.

Excluded people constantly forget that the D2B audience already sees and hears their thoughts, and so they say out loud or act out the sounds they hear or see in their thoughts. But for the D2B audience, it is generally annoying, because it's like seeing or hearing the same thing twice- like an echo. Many images on the excluded thought screen are images that are externally sent there, not that are internally written by the excluded brain. Alternatively, a thought image may be externally triggered to be internally written onto the excluded thought screen (by making you remember some image you actually saw earlier). Probably most images of people, in particular of famous people, that are sent to excluded people's thought screen are identity theft-sent by somebody other than who the image is of. This may be just to annoy the poor excluded people with reminders of people they don't like, or to try to get the excluded to dislike the person in the image, or to get the excluded person's opinions of the person in the image, or simply to advertise the person in the image. Possibly some images sent to excluded might be from D2B consumers that are trying to get your current opinions about them. Another possibility is that images sent to excluded people's thought screens, may be to convey some message to an excluded (that is, to use remote neuron reading and writing as a regular telephonic communication system with those who are excluded-which, in my experience is rare). As excluded this makes you realize: how many people have your image on their thought-screen but you don't get to review their thoughts? What important thoughts and news about us are we missing?

Who owns and controls D2B?

So who owns all these microscopic remote control cameras, microphones, and neuron reading and writing devices if such things actually exist? I don't know for sure, but it seems very likely that the telecommunications companies own and operate most, if not all of them. This includes the telephone (both wired and wireless), and cable companies. One argument against this is that neuron reading and writing apparently predates the existence of communications companies (although communication companies have secretly existed far back into the past). Clearly there must have been companies formed to sell D2B service quickly after it was invented. The initial discovery of microscopic microphones and cameras must have been very exciting, and after the initial celebration, there must have been a lot of thought about how to distribute the new technology among the inventors and their friends. Probably, long ago in the past, perhaps even in the 1200s, wealthy science-minded people developed camera, microphone and information recording technology and also, in the process, discovered remote neuron reading and writing. Then there must have been a massive focus on miniaturization. Probably they used their money to form small secret businesses to grow the technology, but all secretly- just among their wealthy friends. They do the science, and then they and wealthy investors use their money to fund the labor involved in growing and mass producing the microscopic camera and D2BW technology. Those early inventors and developers of micro-microphones, communicators, cameras, and neuron readers and writers must have enjoyed sharing video and sound recordings with each other (probably many of nude women and sex, but also of secret important truths surrounding crimes, and scientific ideas and technology in their own and other nations), but also it was probably important for them to be in quick communication with each other in case of emergencies.

Another possible RNRAW owner and developer may be governments; in particular the military portions of governments, or perhaps even the communications ("postal", etc.) portions of governments. Perhaps the neuron companies and developers are divided by national and even smaller territorial boundaries.

Whoever it is that owns and controls most of RNRAW and D2BW, they must have all the best answers to any questions humans of Earth might ever ask, because of the many years of secret scientific research that resulted in the many invisible secret nanodevices that we feel the effects of everyday. Unlike their poor excluded victims in the public, the D2BW owners are probably very difficult to trick and deceive; they probably don't waste a second of time entertaining all the far-fetched theories the poor public is bombarded with and wrongly believes (like the claims of religions, that 19 hijackers did 9/11, Oswald killed JFK, the universe is expanding, that time contraction and/or dilation can occur, and that light is a massless electromagnetic wave).

It seems likely that there are at least two major Neuron groups, just like there are two big political parties in every nation – and probably the division is along traditional lines of non-religious vs. religious, conservative versus liberal, majority rule versus minority rule, etc.

Another important question is: Throughout history have communication companies recorded copies of all the messages they delivered? Perhaps telegraph companies secretly kept copies of telegraph messages. They must have known the value of knowing what important people are saying to each other. Once a message is in electronic form, making copies of the message can be done very discretely.

Then the secret information can be sold to other people to get more money. It seems very likely to me that the phone companies (and now the cable companies too) have recorded every phone call ever made on all the wires that they own. Wouldn't it be fun to hear famous phone calls? It would be fun to hear our first phone call too. That the phone company probably does record many phone calls and Internet information, implies that they also record thought image and thought sound communications. If true, then the phone companies' archive of data must make the Library of Congress look small. Alternatively, if some part of government is involved, perhaps the actual Library of Congress is much larger than the public realizes, or perhaps there are very massive military data archives that the public, who funded them, does not have access to. It seems very likely that many images of people inside their houses, and their thought images and sounds are recorded and permanently stored by the RNRAW owners and controllers. I, for one, certainly hope that every single neuron writing transaction has been stored, because many of those unpleasant remote neuron writings need to be "paid-back" onto the neurons of the sender.

Have the owners of neuron communication technology and neuron segregation had an undemocratic and unfair influence over the representative-democracy governments of Earth? Certainly by excluding many millions of people from access to D2BW, they are restricting democratic voting, because it's difficult to count the vote of somebody who can't see or hear what is being voted on. Many D2B excluded form very inaccurate views about politicians, because the excluded don't see all the violence that the politicians secretly support, like 9/11, the Kennedy killings, etc. A good example is how, as a 19-year old excluded kid, in 1988, I actually voted for George Bush Senior, not knowing that he was so closely linked to the murder of JFK, and probably the murder of many innocent nonreligious people. If the future of government is full and constant democracy where people vote directly on the laws and decisions they must live under, then you can understand the importance of having a transparent and democratic neuron communication service where the public can see if there is any dishonesty or secret agreements. It must be a natural system for some wealthy people to want to horde resources for themselves and leave the poor as uninformed and uneducated as possible. Hitler and Pol Pot were two of probably many typical examples of popular governmental leaders who may have been hand-picked puppets of RNRAW ultra-wealthy owners that prefer to leave the public in ignorance and to trick them with big lies.

Because all of remote neuron reading and writing has been secret for 700, or however many, years, imagine how many humans who labored to develop and improve the remote neuron reading and writing technology are completely unknown to the public. Of course this list of people also includes many unknown remote particle heroes and villains.

If you think that the public should know about the history of science, about remote neuron reading and writing, and direct-to-brain windows then you are an instant enemy of those people who currently monopolize RNRAW, and they use their extreme technological advantage against you. It's helpful to try to put yourself in their position, and see the universe from their eyes. Each new generation of D2B owners faces the same situation. They are all born on a runaway train of D2B excludocide of much of the human race through denial of providing them with direct-to-brain communications. So you can see why each generation of D2B owners would fear going public with RNRAW and D2BW, not only because it increases the competition for the best resources, but because they might be blamed and punished for centuries of abuses done by generations before them when many people see how they (and those before them) kept RNRAW and

D2BW all to themselves. The feeling that the people of Earth would look at them like they are heroes for going public with D2B is probably out-weighed by the fear that they would be viewed as scoundrels for keeping RNRAW and D2BW for themselves for such an extraordinarily long time. But certainly, they must know the public mind and human behavior so much better than we as excluded do. In addition, the current generation of D2B owners ever losing control over RNRAW and D2BW seems unlikely, but as excluded we can't be sure.

Without question the controllers of remote neuron reading and writing technology have very low ethics compared to average people. This is the only explanation that can account for the way the most obvious truths are mysteriously and secretly denied-raising thought-audios like "...gee, was JFK shot from the front?...how could Sirhan cause powder burns on the back of RFK? ... could light be made of material particles? ... might remote neuron reading and writing be a valuable science?... how could three steel frame buildings fall into dust from fire?...could that 1936 red-shift be from the different sizes of the spectra? ...", and so on with the most basic and obvious truths that are forbidden by each successive generation of D2B owners. Because the D2B owners and operators are so terrible, the D2B consumers so addicted and scared, and the D2B excluded so under-informed, these ridiculously simple observations continue to fail to reach the majority.

The picture in my mind that seems to be confirmed all the time, is that the neuron owners, even those on the "good half", are not a group of gentle-men, not intellectuals, not a sensitive people. But instead the view is that they are a pretty crude, low-brow, anti-education, greedy, very selfish, overly-tough, dishonest, group of people that have very obviously slanted and unreasonable views. But, to their credit, I have to admit that I am still alive to type this and you still alive to read this.

Life under the Neuron is very similar to life under a "Cam-munist" and "Not-see" government

There are a lot of similarities between the current system of life "under the Neuron" and life in nations that live under a single life-long all-powerful supreme leader and dictator like North Korea. How much are the neuron owners like a "Kim Jong-il-Neuron" and son "Kim Jong Un-Neuron"? Just like in North Korea, the neuron owners are a singular, all-powerful, life-long, central dictator. Just like North Korean medal and award winners thank the "supreme leader Kim Jong-Un", so medal and award winners under the Neuron "appreciate it". Those under the Neuron must be even more humble- not even thanking "supreme leader Neuron" proudly, openly, and vocally, but only indirectly through praise phrases like "I appreciate it" and "keep in mind". Anybody who objects is sent away to the virtual gulag of "excluded" for large periods of their life. Anybody who tells the public the truth about remote neuron reading and writing and direct-to-brain windows is labeled a "rat". Any remote neuron writer who sends an excluded person unauthorized D2B windows is no doubt harshly punished with exclusion, loss of job, loss of D2BW, and probably lots of remote molestation sent by their former fellow remote particle "capos". People cannot even admit openly and publicly that they receive direct-to-brain windows, and that they can see and hear thought. Not only are they strictly forbidden by the supreme leader from admitting that they get D2BW, but cannot even be silent or say "I can't say" if ever asked - they absolutely *must lie* and say "no". Many D2B consumers most certainly must take it even further forcing those who ask if they "get direct-to-brain windows" to undergo counseling and psychiatric detention. D2B consumers are intimidated and coerced to vocally lie about seeing and hearing thought even to their children, parents, and friends. They are often forced, out of the blue, to make a clear "pledge" or "oath" statement out loud by stating "I don't think people can hear thoughts", etc., as if they were suddenly required to clearly show subservience and obedience to the Neuron or are making some kind of "opposite" statement as a protest for the audio record. D2B consumers know better than to even think criticisms of the supreme neuron leaders and the system of neuron D2BW secrecy.

How about the similarities to the big lies told under dictators like Adolf Hitler and Pol Pot? Nobody can talk publicly about remote neuron reading and writing as an actual scientific field, nobody can admit that light is made of particles that are material, and that light particles are the basis of all matter. The ridiculous non-Euclidean theory, and theory of relativity reign unquestioned with thousands of multi-million dollar "big lie theorists" constantly publishing page after page of text on the latest radical forces and particles of nature. The obvious fraud of the red-shift, as shown in the Humason 1936 image (see my book "Big Science Secrets, Lies, and Mistakes"), shows not only the long-lived nature of the filthy lies under the neuron, but how even obvious truths can be easily suppressed from the public, and how easily all scientists and teachers can be made to echo the neuron party lie. Imagine yourself as a school teacher having to teach young people dishonest scientific theories that, as a D2B consumer, you know are not true. For a compassionate thinking person who cares about young people and education that must feel terrible. Beyond scientific lies, are the lies involving violent crimes, like the "19-hijackers brought down the 3 WTC buildings" and "Oswald shot JFK" lies, which are exactly parallel to the big lies of the worst dictators of history.

Brilliant excluded people telling the truth are remotely watched by millions of unseen people all day and night, producing billions of dollars for the supreme leader Neuron, but are dismissed as "kooks", "perverts", and "freaks", not receiving a penny of those billions they generate from the millions of one-way-only D2B voyeur viewers. Those "most watched" "kooks", "perverts", and "freaks" are never called what they actually are, which are "heroes", "popular democratic leaders", "very honest", "interesting", and "bright" humans.

There are many clear parallels between the current Neuron D2B system and the life-long monarchies of history. The Neuron is so powerful, that this comparison has never even been made publicly until now. Even this book is barely public and certainly will not be in wide distribution for a long time if ever.

Like any monstrous and murderous dictator, you don't have to look far to see all the victims of the Neuron. The thousands of WTC victims, the Kennedy's, Lennon, millions of galvanized victims-probably one of your friends or relatives was galvanized by old supreme leader "Kim Jong" Neuron, or will be.

Probably for any human born into great wealth and ownership of direct-to-brain remote neuron reading and writing technology, the natural tendency would be to stay with the same system of hording everything for themself, and to not provide any D2BW to other humans; letting other people see and hear thought could only threaten their control over the public and over D2BW technology. But it seems likely that opening D2B to more people would increase the income of the system. Going public with D2BW would also add more variety to the people that can currently communicate through D2BW. Instead of a few people hogging seeing and hearing thought and neuron reading and writing technology all to themselves for centuries, isn't a better system letting everybody talk openly about D2BW, going public with it, and letting the majority control it?

Should a few individuals like AT&T's Randall "Kim Jong-il" Stephenson, who rule for life until their son Randall "Kim Jong-un" Stephenson takes over, be the people who get to decide which "Steve" gets to have "an in son", and which "Steve" lives a life of childless exclusion?

Is putting a singular, group of AT&T (and Verizon) "Neuron" families, with life-long rule, that hordes and controls who gets to see and hear thought, and that decides who gets to receive D2BW and who doesn't, the best system for a nation's and planet's communication service? Isn't a much fairer system a democratic system where the public decides who gets to see and hear thought, and who doesn't?

We can feel good about a few of us being able to know about, and talk somewhat openly about this terrible truth about the unseen Neuron pyramid scheme. But we have to accept that the current situation is terrible. You can see the potential future: there may be many more dismal centuries before the public finally even recognizes that RNRAW exists and has been secretly in use for centuries, then two or three hundred years more before talking about D2BW segregation is widespread and is a common discussion, and then a few more hundred years before most of the public gets even a very limited form of D2BW.

Are there any tips to getting included?

There isn't a lot I can say here other than: try to be in good physical shape, don't commit any crimes, and don't bite on suggestions that will make you look bad or do something unpopular. Try to fit in by understanding that the D2B consumers can see and hear your thoughts. Outside of that it's not clear what might help.

One question is: does talking openly about RNRAW and D2BW help, hurt, or have no effect on an excluded getting included? Some of your D2B "friends" will probably take a few thousand dollars to make a hand or finger over mouth gesture to you, "advising" you, as an honest and trusted buddy: "shut-up about D2B and you will get included just like me". But they get a lot of money to lie and mislead you, in particular if you trust them. Personally, I think we excluded are much better off talking about it, and trying to inform the excluded public. The D2B owners and consumers probably all know that 99% of those who are included stop talking about RNRAW once they are included. Probably most are older and simply try to pick up the pieces of their ruined lives, try to find a mate and make a family, etc. They probably quickly realize that an individual means nothing and is powerless against the massive all-powerful neuron industry. One theory is that, if you say nothing publicly about RNRAW and D2BW then you are "not a problem" (a NAP) for the Neuron, so they will probably ignore you and leave you as excluded for the rest of your life. But if you do talk a lot about it, then you are a problem for the Neuron, and they may feel compelled to try to silence you. Of course that could take the form of violent crime, and definitely does take the form of intimidation and threats of violence, but probably more likely it could take the form of trying to buy your silence with some kind of agreement. It may be that you fit into one of the two sides, as either a conservative or liberal, and so each side simply "includes" like-minded people at some time which they determine by majority vote of D2B consumers that receive their service. As an excluded who knows about D2B you probably will find that your D2B "friends" constantly say some word that is supposed to indicate when you will be included. But, knowing how many people are excluded, about Thane Cesar, 9/11, the expanding universe fraud- it seems more likely that those who control D2B mostly earn the nickname "inhumans". It seems likely that, despite all the D2B consumer promises of "we're going to get you", etc., that most people are excluded for life, because notice that the D2B *does not negotiate* with individual excluded people, for example, provide you with some limited D2BW service, offers of money, chat with you through D2B, etc. in exchange for you to stop talking about D2BW. My view is that they need to show me some regular D2B service before I will even think about offering anything in return. If you don't talk about D2B, and are excluded for the rest of your life, which is likely, then you did nothing to help end the segregation not only for yourself but for future generations of victims.

One possible truth may be that getting direct-to-brain windows may depend in some part on what "air-space" you live in, because it may be that the air around you is filled, of course, with nanometer size particle devices, but the majority of those devices, and nano-device space supremacy, may belong to the prevailing political party. So for example, if you are a liberal living in a conservative place, your chances of remaining excluded are probably higher, and the same may be true if you are a conservative living in a liberal place.

The initiation to D2BW process; when the D2B "gets" (includes) somebody

Clearly many humans get D2BW. So at some point in their life they must have the service turned on. What is the process for a person who is newly initiated to D2BW? Perhaps there are just more and more stronger and longer lasting D2B windows over time- perhaps starting with those during and waking up from sleep. I doubt that a video window just pops open directly in their eyes, and some guy starts talking directly into their ears. Perhaps, for many people, a D2B consumer, maybe a family member, sits down with them and goes over some well-worn script used to include people, like "you've reached an age where you are ready to handle more responsibility…". Maybe there is an analogy for why the newly included person must keep everything about D2BW completely secret, from those excluded, but even from other D2B consumers. For example, they may use the example of homeless people; that there will always be people who are less fortunate than others, to help create reasons for the brutal secrecy. They must be very clear about the need for secrecy. Perhaps there is a very scary story, maybe about violence done to those who did not keep the secret. Then perhaps there is a slow introduction, like a thought-audio message from the D2B consumer sent to the ear of the newly included like: "…do you hear my voice?...". Probably many times the newly included person says "yes" out loud, so the next line must be (again in thought-audio) "ok but the first rule to remember is that we can't answer out loud. Can you answer my question just in thought?". Then they probably introduce D2B windows. Probably there is a thought-audio statement like "you are about to receive a picture, directly to your eyes for a few seconds. It may seem uncomfortable at first, and you may feel that the picture is too close to you. You may feel uneasy because the picture moves with your head and you cannot turn it off. ...". Of course, I can only guess, but there must be some well worn introductory scripts. Similar scripts may be used initially with the public by a D2B device manufacturer or by government employees once D2BW goes public. There is a funny analogy with the first motion pictures shown to the public in theaters- one movie was of a train- and the public who had never seen motion pictures always leaned out of the way when the train came toward the screen.

Is there perhaps a lack of D2B resources that prevents most people from being able to get D2B?

The answer to this seems most likely to be "no". By the year 2000, like the Internet, the D2B technology must basically permeate the planet Earth and beyond. Seven billion people, the current human population of Earth, is not perhaps as large as many people might think. If each person was a byte of memory, they would only total 7 gigabytes; which is small by today's standards. But, certainly, there must be a limit on how much data can be stored, and for how long. The movie "The Final Cut" hints that only the most important movies of most average people's lives are permanently stored after their death- many are probably deleted to make space for those still living.

But in terms of direct-to-brain windows, there is no reason to deny the public this basic service. We all probably have RFID neuron organelles in our neurons, as crazy as that may sound, and are "D2B-ready", but the D2B service is simply not "turned on" for us. Instead, D2B is used to murder, assault, molest, and mislead millions of people.

Will there ever be a "big pay back" or "big turnaround" where the murderers fall and their victims rise to the top?

This is a question that I'm sure many excluded who realize that they are being denied seeing, hearing, and communicating by thought think about. Will there ever be a big payback? I don't know. Without a doubt, the possibility of excluded people themselves doing much is very low, because of the extreme and absolute advantage RNRAW gives to those who own and control it. Any reversal could only occur at the top: by neuron owners, and/or possibly by D2B consumers. But probably we excluded at the bottom can contribute a tiny influence over long periods of time.

There are historical examples of similar situations. One example is that of slavery and women's rights; clearly slaves were freed and both the freed slaves and women eventually won the right to vote, but there was no fast reversal or pay back for the centuries of injustice. But other examples include the US, French, and Russian revolutions where the wealthier subjects, to a large extent, overthrew the old rulers and became the new rulers. So it's tough to really know. If I had to guess, I would say that it's going to be a slow and nonviolent progress into the future where more and more excluded get included and more D2B consumers start talking publicly about D2B. Of course, for many excluded the dream may be for a fast reversal: the D2B and 9/11 killers jailed, all the lies and liars exposed, all the victims compensated, and those who told the truth richly rewarded. There are cases where the sides have quickly reversed, like WW2, etc. Clearly it would be a case, in the USA, and probably all over the Earth, of the murderers (of the Kennedys, 9/11 victims, etc.) and liars being on the losing side, with all the victims of murder, assault, and remote molesting and their allies being on the winning and ruling side – controlling remote neuron reading and writing, showing all the thought-images and sounds of the past to the public, exposing and jailing the many murderers that are currently being protected by the neuron lie and secrecy, turning on D2B for the many excluded, and freeing the many wrongly imprisoned.

One wonderful aspect of any kind of reversal or turnaround is that the public will quickly see and know, for sure, who is a killer, who fought against the killers, who lied, and who told the truth. Even as an excluded, this can be seen simply by recognizing those who mislead the public about 9/11, and all the other big obvious lies. Perhaps the newly informed public might seek to reward all the new heroes: those who fought against the violent, who told the truth about 9/11, remote neuron reading and writing, and science, and who turned down the money to remotely murder, assault, molest, lie, and shill – and seek to punish all the newly revealed scoundrels: those who did 9/11, that remotely murdered, assaulted, molested, etc. and that actively accessorized to try and trick the excluded from knowing the truth of many murders and lies. RNRAW and nano-cameras going public would be the end of many big lies and tricks because tricking and lying to people would be very difficult when all thought-images and thought-sounds are freely and widely shared.

One terrible problem is "forgiveness", as harsh as that may sound, but because forgiving is a central theme of Christianity, there has been a long tradition of forgiving those who murder. In my experience, forgiving and "turning the other cheek" is uniformly interpreted by the evil other side strictly as weakness and as an opportunity for more murders. The smart machines that the particle murderers use recognize this reality. We owe it to the many millions of victims, many of whom are not alive to complain, to make sure that their murderers are punished and all the videos of crimes made available to the public.

There may be an age, sometime in the near future, of many "big take-downs", where hundreds of people with many counts of remote particle murder, assault, and molestation are finally identified, removed from the particle technology, and jailed.

Each particle crime probably has an electronic record. Each murder, assault, muscle move, particle molestation, or money-window shill, are all transactions that must pass through the telecom companies' networks. Presumably each of those transactions is recorded and stored. If true, which seems likely, then that data, if it survives the transition from secrecy to being made public, can be used to punish those who have not yet been punished for their particle crimes, and for those people who committed crimes to be forced to pay back their victims. Such records must clearly show who are the particle murderers, assaulters and molesters, who paid them, how much was paid, in addition to who were the victims. Then clearly, all the criminals should be fairly punished, and all money involved given to each particular victim. I think most people can agree on punishing violent crimes, and compensating the victims with all money that was involved. In addition, all money gained for nonviolent remote molestation, and even for "shilling" may be voted to be paid to each particular excluded victim.

There are a lot of possibilities of a future where a majority of people try to "right" the "wrongs" done to D2B victims. One example of this is that the majority may vote to end the D2B read and write service for: those who voted "no" to going public with D2BW, for those with first strike violent crimes, and for those with multiple remote assaults or molestations.

Other reasonable thoughts and ideas about a future reversal include:

1) A fine for each proven remote molestation that equals all the money paid and at least $100 per second of molestation which must

be paid from the molesters directly to the victim, in addition to the molestation being replayed on all involved in the funding, ordering and sending of the remote molestation.

2) At least 1% of all money paid to the neuron owners to see videos of a person or their thoughts should be paid directly to the person being watched.
3) Democratize at least one remote neuron read and write service for the public. Private people can make a competing neuron reading and writing service, but at least one should be majority rule, so that people who are Democrats, Green party, Libertarians, etc., are not excluded from using remote neuron reading and writing, simply because Republicans own the only neuron companies.

Basic wants for the future

Some basic wants for the future are to:

1) Stop the remote molesting (muscle moves, itches, unwanted thought-sounds, etc.). Let people see "who wrote it" to their brain, remove the neuron write permission from those who remotely molest, fine them, and give the fine to the victim.
2) Go public with D2B technology, and demonstrate it publicly. Then everybody may not receive D2B but at least they can see that it exists, what it looks like, and how it works with many video examples.
3) Remove FSV (first strike violent) people from access to remote particle devices.
4) "get out"- educated D2B owners and consumers should focus on including more people that are currently excluded- in particular the non-religious, evolutionists, anti-violent, pro-racial and gender integration, pro-consensual sex, educated, pro-democracy, those who have no problem with free-flow of all information and loss of privacy and secrecy, etc.)

Perhaps there could be a large law suit against the telecom companies, in particular AT&T. They clearly have broken many laws in their use of particle technology. Obviously the most serious are the violent crimes: **homicide**, **assault**, **conspiracy to commit homicide**, **accessory to homicide**, and then the non-violent crimes: **molestation** (done remotely on motor and sensory neurons with particle devices, which is still covered under the molestation laws and so is illegal presumably). Clearly the telecoms are violating the **privacy laws** in using micro and/or nano technology to capture and sell images and sounds from inside people's houses. Many nations have laws that clearly prevent other people, in particular non-governmental people, from capturing and selling images and sounds of people in their own homes. But there is no law, that I am aware of that forces people to compensate the subjects for images and sounds captured and sold of them. Another clear set of laws being broken by the telecoms are the **copyright laws**. Presumably all works made by any person are copyrighted, and the telecoms freely copy and distribute any and all works of people to D2B consumers, for money, without getting the permission of, or compensating, the owners of all those copyrighted works.

But beyond the clearly illegal activities, there is the massive discrimination being done against the "D2B excluded", which may not be illegal, and which has obscure but very real and enormous consequences that include inability to reproduce, discrimination in hiring, and the other abuses that come from many people viewing one person without that person having the opportunity to view those people watching them. Most likely any kind of law suit would be best done as a massive and collective effort by as many people as possible to try and make a permanent change of making D2BW available to the public (as available as cell phone service) and to compensate as many remote particle victims as possible. Beyond a law suit, perhaps ballot measures can be passed by the public, or new laws enacted by government representatives, to open and start to democratize direct-to-brain windows, and to identify and compensate the many victims.

When do you think D2BW and RNRAW will go public?

It seems likely that within the next century, and perhaps even within this or the next decade, remotely hearing what the ear hears and even hearing thought-audio will be made public. One of the authors of the 2008 paper which published the first public image of remote neuron reading (what the eyes see)[4], Yuki Kamitani, told me in an email message in 2009 that they already can distinguish between different syllables of thought-audio with magnetic resonance imaging, so in some sense hearing some aspect of thought-audio is already public knowledge. But also, the first public thought-image may be published within the next 20 years, probably by people in a University in Japan, or at the University of California at Berkeley. But from there it will be many years until it reaches the public in the form of handheld cameras that capture images and sounds of thought, and perhaps by then, D2BW will be available to the public.

One possible rule is that if one or more excluded people have figured something out, and are making some product or project, this is a signal to D2B owners that the technology or project the excluded are developing is "ripe" enough to allow wealthy D2B consumers to go public with the same technology or project on a larger, more well-funded scale. This may serve several purposes, for one thing, it takes the steam out of the excluded product and project-otherwise, the excluded might have a monopoly on the "new" idea or technology and have a large influence over the public, and also, the D2B owners can then step in and control the growth and

[4] Miyawaki, Y., Uchida, H., Yamashita, O., Sato, M., Morito, Y., Tanabe, H. C., Sadato, N., Kamitani, Y. (2008). "Visual image reconstruction from human brain activity using a combination of multi-scale local image decoders. ", Neuron, 60, 5, 915-929. http://www.cell.com/neuron/abstract/S0896-6273(08)00958-6

distribution of the "new" technology or project. Some good examples are: walking robots (both with motors and artificial muscles), devices that can record sounds ears hear and thought audio, devices that capture images of what eyes see and thought images, intracellular wireless devices, microscopic flying wireless cameras, ships into orbit, and movies that tell the story of evolution, science and the future. Excluded people are slowly starting to use the available public technology to build walking robots, to hear thought, to build ships that can get into orbit, to make movies about evolution, etc. So, it seems very likely that the D2B owners may allow D2B consumers to step in, and capitalize or "cash in" on these kinds of projects before any excluded people might take the initial public thrill and excitement from the deprived excluded public when finally the "new" products and truths do reach them. For this reason, many D2B owners and consumers may write on talented excluded to try and force the technology and truths out to the public to gain D2B approval to go public on a larger scale. Either way, we all benefit- although the raw and hard-earned truths and technologies the poor excluded, as "worker bees", "reinvent" are left to history as relics of a dark age, they stimulate and force similar and much larger scale improvements to reach the larger excluded public. But wouldn't a much better system be for the D2B owners to lead the way into the future, instead of waiting for excluded people to "reinvent the wheel", before they take even a tiny step forward?

Knowing about remote neuron reading and writing is a huge advantage

As a person who now knows something about remote neuron reading and writing, you have a tremendous advantage over those poor people who know nothing about it. Just knowing about remote neuron reading and writing probably increases your chances of survival, of reproducing, of getting physical pleasure, of getting employed, of finding housing, etc. As a person who now knows about D2BW and RNRAW, but is probably a D2B excluded, you may now be viewed by many people as an "excluded who knows", which may be a step above the many other people who are marked as simply "excluded" (but still lower than a "D2B consumer"). The main advantage is that you probably will not bite on the many bad neuron written suggestions. But still, your chances for a date, a job, and reproduction are extremely low.

Once an excluded person knows about RNRAW, getting them to "bite" (repeat/act out) bad remotely written suggestions is much more difficult- it must be a similar phenomenon to remotely writing onto the brains of D2B consumers- because they know to reject violent and other inappropriate suggestions. But even so, there can still be "remote control humans". Does anybody who understands RNRAW doubt for a minute that, like our mouth or finger muscles are routinely moved for a second, that our entire bodies could not be remotely controlled indefinitely down to the last neuron? This truth creates a constant "Invasion of the Body Snatchers" phenomenon and fear in people, because like that classic horror movie, nobody can ever really be sure that their friend, mate, or even they will not be suddenly remotely controlled by unseen criminals of the opposite side, to involuntarily do violence or other unpleasant actions. I know from experience that to have a positive moment with a friend, instantly prompts violent and inappropriate images and suggestions onto our minds, always requiring an instant "fire back" in thought at the enemy, as odd as that sounds to those not familiar with RNRAW. When I speak or play music publicly, I have to constantly fire back in my mind at remote molesters- much of my life, and no doubt the lives of many other people, has been an endless "fire-back in thought at remote molesters" movie.

Just to give you an idea of how shockingly terrible the time we live in is, and in particular how far we have to go in terms of teaching the public about remote neuron reading and writing, I can say that, there was no phrase "remote neuron reading and writing" until I started using it on my web page, certainly by February 2010. Nor was the phrase "direct to brain windows", which I started using certainly by April 2011, ever stated out loud or published, to my knowledge, anywhere, publicly before. If you search the Internet for "remote neuron reading and writing" and "direct-to-brain windows" you will probably only find my web pages; even now in 2012, there is no public acknowledgement of the importance and secret of remote neuron reading and writing, or the massive secret society of those who receive direct-to-brain windows. How did I figure it out? It's possible that it was remotely neuron written onto my brain. I probably have received helpful remotely written hints (in between the endless remotely written garbage). As I said, I've never seen the phrase "remote neuron reading or writing" in print or heard it out loud ever in my life, and I have scoured the history of science, looking in particular at many physiological papers dealing with neurons. My best answer is that I haven't been handicapped by any religion, and also, in particular, that I have spent the last 8 years researching the history of science for my "ULSF" project. So this gave me an unusually good perspective on the course of science history as experienced and recorded by many of the great scientists of the past.

So, even now, to the best of my knowledge, and no doubt, for many years to come, there does not exist anywhere in print anything about remote neuron reading or writing or direct-to-brain windows. In fact, I don't know for sure what those who own and receive direct-to-brain windows even call it. I've called remote neuron reading devices "brain imaging machines". Another acronym I made is "SHASIASTAFB" (Seeing, hearing, and sending images and sounds to and from brains). Traditionally RNRAW has been called "telepathy", and now remote neuron reading is being done with "functional magnetic resonance imaging" (fMRI). It can also be called "remote neuron activation", and "direct-to-brain videos", or perhaps some other thing altogether. The phrase "direct-to-eyes" may sound less intrusive and crazy. Recently on Wikipedia I found a page on a "virtual retinal display" http://en.wikipedia.org/wiki/Virtual_retinal_display (basically an image sent "direct-to-retina") which is similar to D2B. An interesting analogy between the human brain and another electronic memory – a computer memory- that arises from the realization about remote neuron reading and writing, is the secret field of future technology called "remote transistor reading and writing" where even a tiny semiconductor transistor in a computer or robot somewhere can be remotely read from and written to using tiny nanometer flying or floating particle device networks.

Why do so few excluded people realize that they are excluded from D2B?

This is really unbelievable. The only answer I can give is that 1) they have never heard about remote neuron reading and writing- they know nothing about the scientific history and all the hinting, 2) they don't want other people to think that they are crazy by saying that they think people see them in their house, or hear their thoughts, and 3) they are remotely neuron written on to dismiss even the hint of such talk as unrealistic and/or unimportant. One constant unfulfilled dream of mine is to find an excluded woman that actually realizes the truth about direct-to-brain windows to talk to about it openly as a friend and possible mate to make a family with, and that a D2B included woman may step over the barrier to talk to me openly about direct-to-brain windows, again, even just as a friend, and, of course, also possibly as a mate. Sadly, this hasn't happen yet, but I'm hopeful and optimistic.

How could a secret like remote neuron reading and writing be kept secret for so long?

One answer is because the remote neuron phenomenon is a mixture of the evil, who murder and lie (9/11 being a fine example), the scared, who are too afraid to tell the excluded public the truth, the stupid, who support the evil murderers, and the under-informed excluded who can't recognize that the evil are murdering and lying. The collapse of science and the rise of the opposite side around 400 AD may explain the prolonged dishonesty, secrecy and idiocy. Well entrenched traditions of dishonesty and fraud have existed for centuries- religions being the main source of many supernatural and dishonest claims (Moses parting the red sea, Jesus turning water into wine, the many claimed "relics", psychics, horoscopes, fortune cookies, etc.). Suppression of science, honesty, and education is a large part of the possibly 700 year delay. Making a muscle contract using electricity dates back (in public documents) to the 1600s, and remotely contracting a muscle to the 1700s, but yet the vast majority of people have never heard about this, and modern scientists have only recently (1959) even gone public with a pacemaker that can remotely fire a neuron. Another example of the lack of public education is that: in 1208 the theory that all matter is made of light was made public, but even today, this theory is not only not publicly accepted, but is not even publicly mentioned as a possibility, for example in the Encyclopedia Britannica. Wars have not helped to bring science to the public, in particular the most recent World Wars, during which most scientific publishing to the public stopped.

Beyond that, there is a strong feeling for many people to "do nothing", and also the feeling of "I get D2B, so who cares about those that don't?". One reason why D2B consumers keep the secret is clear:

because they don't want to lose the extremely valuable ability to see and hear thoughts. In addition, many people prefer an uneven playing field, where they have an advantage for mates and jobs, etc.- a newly included person might get a job or mate that you want. The more people that are included, the more competition there is for finite resources.

Imagine an alternative existence where we were born on a planet where everybody receives D2BW and talks about it openly, where everybody knows that all matter is made of material light particles and that our future is to build a star cluster, where everybody has already learned about evolution, where there is no antipleasure ferver, where enlightened full and constant democracy rules, where people routinely travel to other planets and other stars, where walking robots do all the manual labor, where violence and lies never happen: then we wouldn't have to spend endless hours writing books like this trying to tell people the most simple truths only to be looked at like we are crazy.

Shut up or speak up?

For myself, the answer is clearly "speak up". An analogy is: "Teach the public how to read and write, or keep the knowledge of how to read and write to yourself?

Made in the USA
Monee, IL
04 November 2026

67377666R00095